Trends in Enterprise Knowledge Management

Trends in Enterprise Knowledge Management

Edited by
Imed Boughzala
Jean-Louis Ermine

Series Editor
Pierre-Noël Favennec

iSTE

First published in France in 2004 by Hermes Science/Lavoisier entitled "Management des connaissances en entreprise"
First published in Great Britain and the United States in 2006 by ISTE Ltd

ISTE Ltd
6 Fitzroy Square
London W1T 5DX
UK
www.iste.co.uk

ISTE USA
4308 Patrice Road
Newport Beach, CA 92663
USA

First South Asian Edition 2007

Library of Congress Cataloging-in-Publication Data

Management des connaissances en entreprise. English.
Trends in enterprise knowledge management / edited by Imed Boughzala [and] Jean-Louis Ermine.
p. cm.
Includes bibliographical references and index.
ISBN 13: 978-1-905209-03-3
1. Intellectual capital--management. 2. Knowledge management. I. Boughzala, Imed. II. Ermine, Jean-Louis. III. Title
HD53.M354 2006
658.4'038—dc22
2006006216

British Library Cataloguing-in-Publication Data
A CIP record for this book is available from the British Library
ISBN 10: 1-905209-03-7
ISBN 13: 978-1-905209-03-3

Printed in Brijbasi Art Press Ltd., I-72, Sector-9, Noida, U.P. India.

Table of Contents

Preface

Knowledge Management

Today the economy of knowledge is already a reality. It is recognized by the European Union as a major factor in durable growth. Recent studies by the OECD[1] have shown the influence of Knowledge Management on the productivity of companies. Companies constantly test, in a precise or diffuse way, the consequences of actions that touch their knowledge capital, whether in a positive way (if, for example, their know-how leads to an increase in market share), or in a negative way (if, for example, knowledge is lost for one reason or another: reorganization, retirements, etc.).

Knowledge Management (KM) started from a number of basic problems and has now developed to become a real discipline with its own problems, methods and tools. It is a developing discipline that borrows from various fields: economics, management, social sciences, information sciences and technologies, educational studies, etc.

The aims of KM are ambitious: to set up devices (organizations, methods and tools) that develop, in a very extensive sense, the knowledge capital that any social organization accumulates during its life cycle.

The challenge for companies is strategic: productivity, competitiveness, continuity … But there is a long way to go. After an "information revolution" that was carried out almost under duress, will we go forward to a "knowledge revolution" (it would be more relevant to speak about a "revolution through knowledge")? The path is being built step by step, and the directions are all still far from being marked out. There is still much room for creativity, innovation and experimentation.

---

1 Organization for Economic Co-operation and Development, http://www.oecd.org.

Two French KM players have joined together to produce this book.

The GET and INT

The Groupe des Ecoles des Télécommunications (GET) constitutes one pole of reference within the French network of public research in Information and Communication Sciences and Technologies. Its research force includes about 550 full-time-equivalent-researchers, and its scientific activity is likely to be doubled by 2005. The GET is composed of four "high" Engineering and Management Schools, to which three establishments created jointly with universities have been added, in the form of GIE (Economic Grouping of Interest).

Among the GET schools, the National Institute of Telecommunications (INT) is unique: for 25 years, it has been made up of two schools on the same campus: an engineering school (Telecom INT) and a management school (INT Management). This combination has constituted a major asset, because today, more than ever, this double competence has become both essential and natural and is focus of concern of most companies.

It is thus natural that INT is developing along the axis of Knowledge Management, which lies between the management of companies and the science and technology of information and communication.

The Club Gestion des Connaissances: the Knowledge Management Club

The *Club Gestion des Connaissances* is an association of companies founded in 1999. Its founder members were PSA Peugeot Citroen, Microsoft France, Cofinoga and OSIS group (Bull). It includes within its structure organizations of all types and all sizes.

Already members of this club are: industrialists such as PSA Peugeot Citroën and Thales; software producers such as Microsoft and SAP; French state organizations such as the General Direction of the Armament (DGA) and the National Navy; research organizations such as the National Office of Studies and Aerospace Research (ONERA), the National Institute of Research and Safety (INRS), the National Center of Space Studies (CNES), Hydro-Québec (Canada) and Radio-France; service companies such as Cap Gemini Ernst and Young and Cegos; and SMEs (small and medium-sized companies).

The companies and organizations that decided to found the Club Gestion des Connaissances considered that knowledge is economic capital, a factor in productivity, stability and determination of competitive assets. They wished there to be no delay in French and European companies compared to their world competitors in this field.

Its objectives are:

– to develop intersector dialogue and debates between decision makers and experts;

– to help managers to locate their actions within the major evolutions in this strategic field;

– to propose thinktanks, meetings, documents, etc., so that in a very short time rich and global information could be collected;

– to increase the dialogue between all players and to extend networks;

– to provide the members with a set of concepts and operational tools that will helping to implement developments within knowledge management in their organizations.

The club functions through thematic workgroups, which study topics identified as crucial for knowledge management: economic aspects, innovation, continuous progress, competence management, business intelligence, knowledge cartography, etc. Each commission, under a chairman, produces deliverables that are integrated into the club knowledge capital to be shared by all the members. Special events allow members to come together for more informal exchanges.

The Knowledge Management seminar

INT and the Club Gestion des Connaissances joined together to organize a seminar entitled *Trois jours pour faire le point sur le Knowledge Management* (Three days to provide a progress report on Knowledge Management) in April 2003. This seminar stressed the importance of KM as a key factor in success and a lever of growth and innovation. This provided a progress report on the whole of theoretical and practical work in this field over recent years. On one hand, recognized researchers in the field presented concepts, methods and tools of KM. On the other hand, players in the economic/business world shared their experiments and their thoughts by presenting experience feedback, case studies and concrete syntheses.

The contributions and exchanges were rich and relevant. The points of view presented were very varied and characteristic of trends in the field. In order that exchanges on these days would not remain confidential, it was decided to produce a book based on the meeting. One will thus find here the major part of the discussions that took place at this seminar. One can thus position the whole of original French work in the field, work that is interdisciplinary based on two complementary points of view: that of companies and that of researchers.

The original French version of this book, entitled *Management des connaissances en enterprise*, was published in March 2004 by Hermès Lavoisier as part of the Technical and Scientific Collection of Telecommunications (CTST).

The content of this book clearly reflects the subjects tackled during the seminar, which corresponded, as far as we could tell, to the major concerns of the companies starting in KM.

Trends in applied Knowledge Management

This book is a translated and augmented version of the original French book with a more international dimension. It includes contributions from authors in different countries (e.g., the UK, Canada, Holland, Argentina, Brazil). It reflects only the opinions and work of that "knowledge network" of contributors, who are, for the majority, recognized as dynamic actors in that field. The work does not claim to be exhaustive on the subject, which would be rather pretentious in a field that some would like to see marked out already, whereas it is only now emerging.

Introduction

Three introductory chapters are given first. Jean-Louis Ermine gives an overview of KM, from a pragmatic and rather engineering point of view. Every succeeding chapter is more or less a detailed study of what is presented there. Patrick Epingard gives an introduction to KM from an economic point of view. This is the fundamental reason why applied KM is set up in companies. Walter Baets gives an introduction to KM from a rather epistemological point of view. The fundamental hypothesis here is that KM is based on complex system theory, which is the framework of each succeeding chapter (even if it may sometimes be hidden!).

Academic studies

The second part of the book is written by academics. It compiles and analyses many applied KM studies and cases.

Nigel Courtney, Clive Holtham and Chris Hendry analyze the experiences of eight varied organizations in managing their intangible assets in the UK, other EU states and the USA. Jean-Michel Viola and Réal Jacob, giving the example of a large Canadian company, show how knowledge elicitation triggers a learning process and tacit knowledge creation. Aurélie Dudezert gives an overview of the different methods and approaches used in companies to value KM performance. Imed Boughzala shows how inter-company co-operative information systems may support KM in the framework of the extended enterprise. Thierno Tounkara puts forward a proposal to link KM and environment scanning, which is a frequent question but one with few answers. Finally, Pierre Fayard gives a very different point of view on KM, inspired by the Japanese concept of "*Ba*", returning to the sources of KM, which is greatly influenced by the Japanese School led by Professor Nonaka.

French KM Club studies

The third part of the book is an overview of the pragmatic work of the companies of the French KM Club and tools that have been designed and are currently applied in these companies and some others.

Jean-François Tendron describes a general evaluation tool, called the "Knowledge Maturity Model", dedicated to revealing the capacity of a company in managing its knowledge capital. Gérard Aubertin proposes a methodology for elaborating a knowledge map of a company and a strategic assessment of each identified knowledge domain, using "Critical Knowledge Factors". Jean-Marie Bézard describes a tool dedicated to evaluating the capacity of a company to be innovative ("Innovation Maturity Model"). Olivier Lepretre, with the tool "Technology Maturity Model", analyzes how information and communication technologies are used in a company to fulfill KM requirements.

Case studies

The last part of this book gives a number of case studies

Nathalie Le Bris describes KM experience in a European automotive industry. Cécile Decamps and Michel Galinier describe how Thales, a worldwide defence industry, organises "Communities of Practice" for KM. The last two case studies are in the nuclear field, which is particularly involved in KM, a top priority in the world for that domain: the first is by Rita Izabel Ricciardi and Antonio Carlos de Oliveira Barroso concerning a radio-pharmacy centre in Brazil, while the second by Marta Eppenstein concerns a nuclear power plant in Argentina.

Conclusion

We hope, with these contributions to show both that KM is alive and concerns a wide range of issues, and that interaction between companies and researchers is fertile. This is because it is true, in this field, that real problems of research are born out of real problems in the real world.

Imed Boughzala, Jean-Louis Ermine

PART 1

Introduction

Chapter 1

Introduction to Knowledge Management

1.1. Introduction

1.1.1. *Knowledge: a strategic value for the firm*

In the last few years, with sudden and unexpected strength, Knowledge Management (KM) has emerged as a major challenge in many enterprises. A set of strategic elements has contributed to this emergence; for example:

– knowledge is a fundamental economical asset ([Edvinson 99], [Strassmann 98], [Stewart 97], [Foray 00]);

– knowledge is a strategic resource [OECD 96];

– knowledge is a factor in the stability of the enterprise;

– knowledge brings a decisive competitive advantage.

1.1.2. *Objectives of KM*

The strategic vision that leaders of an enterprise can have concerning their knowledge capital is leading them to define some global objectives for managing and optimizing this resource. These objectives are always articulated around three key points:

– capitalize: know from where we come, so that we know where we are, to know better where we are going;

Chapter written by Jean-Louis ERMINE, Institut National des Télécommunications.

– share: switch from individual intelligence to collective intelligence;

– create: create and innovate to survive.

Note that these objectives of management are indeed "paradox injunctions" ("Knowledge is Power"!) and imply that KM is an important challenge for managers which involves long-term change and new visions of the enterprise.

1.1.3. *KM: a new and complex approach*

KM is henceforth part of the real life of the firm. But to understand what the management of knowledge in an organization concerns in reality is not easy, because nearly all components of the firm are involved.

KM involves strategy, because it is a new type of management responding to a new socio-economic environment and a new vision of the organization. It involves the structure of the organization because the knowledge is created through complex networks, connected to the environment that can challenge the classic structures. It concerns many processes that have already been put in place in organizations (fortunately, human beings have managed their knowledge for a long time!), but that have to be reviewed from new perspectives, optimized or developed. It concerns the staff of the organization, the human resources, which are the heart of the issue, because knowledge is created, is shared and evolves only through people, who must mobilize themselves personally and collectively for this objective. It concerns information and communication technologies (ICT), which are powerful vectors for KM if they are used efficiently.

In the following, we will attempt to give a set of points of view that will allow a better understanding of these issues.

1.2. The two types of approaches to KM

Knowledge capital in the enterprise is intangible capital, which is not visible in the organization; even its content remains elusive. It can be agreed now that the content of such capital is both hidden in and scattered amongst two essential components of the enterprise:

– *Human and social capital:* the quintessence of the knowledge in the enterprise ("core knowledge") is in the heads of its employees. This is in any case its ultimate place before it is used operationally. It is in a tacit form, and hardly (or even not at all) expressible, according to the adage: "we know more than we can say". This tacit knowledge capital, tightly linked to the human capital, is at any one time both collective and individual and develops through the knowledge networks of the organization ("knowledge workers"), who constantly produce and use this precious and operational knowledge.

– *Information capital:* enterprises have stored, for decades, huge amounts of information, which they stock and distribute with more and more sophisticated methods. Employees of the enterprise constantly use this gigantic system of information for their operational activities; they acquire information and give it a precise sense within their operational context so as to transform it into knowledge that is useful to their profession. Unfortunately, for the right person to find the right information at the right moment becomes a difficult task, when the available amount of information is considered. The potential knowledge is buried in masses of information and is not easily accessible.

Therefore, managing knowledge capital is problematic, because of its two-headed character, with the two types of capital, human and information, and the hidden, tacit or buried character.

To approach this issue, two main approaches are possible:

– the first is to address (at least partially) the hidden character of knowledge by making it explicit, using either the tacit knowledge in knowledge networks or knowledge buried in information systems;

– the second is to retain the hidden character (especially the tacit aspect) of the available knowledge and to managing the knowledge networks.

We are going to examine briefly how these two approaches can be put into effect.

1.2.1. *Knowledge elicitation*

Depending on whether one is interested in tacit knowledge in networks in the organization or knowledge buried in information systems, elicitation of the knowledge capital can be based on two distinct approaches.

1.2.1.1. *Tacit knowledge elicitation*

1.2.1.1.1. Knowledge transcription

The first type of approach is what we will call knowledge transcription: some tacit knowledge can be elicited simply by transcribing it in a more or less structured manner. This is the case in setting up quality systems (of which the first rule is "record what you have to do"), in files reporting experience, or in the producing publications. Furthermore, "secondary documents" can be used to synthesize knowledge contained in the primary documents.

1.2.1.1.2. Knowledge engineering

Knowledge engineering is a more sophisticated approach than transcription to capturing parcels of tacit knowledge. It was developed along with expert systems (or knowledge-based systems). These systems were intended to replicate the reasoning of experts on specific knowledge domains. It was quickly perceived that if powerful technologies were available to design such systems, the critical difficulty resided in the capacity to transfer the knowledge of one or several human experts into a computer program. Knowledge engineering therefore put methods in place to collect the knowledge, most often from interviews, and to structure it generally on the basis of models ([Rubiello 97], [Dieng 00]).

These methods can therefore be used profitably to clarify, from interviews with knowledge holders, a part of the capital of tacit knowledge of the organization. A typical example of this evolution is the MASK method that was used to specify expert systems and became a method of knowledge capitalization, which was then integrated into the general KM approach ([Ermine 02]).

1.2.1.2. *Knowledge extraction*

1.2.1.2.1. Knowledge extraction from data

All enterprises retain large quantities of data, resulting from their production activities. These data are very diverse (technical, management, marketing, etc.) and their mass doesn't stop growing (it doubles every 20 months on average). Moreover, there are other considerable masses of data called non-structured or semi-structured data, which consist of textual data (and other types of media) that correspond to the production of texts, cards, reports and other types of document. This informational capital probably provides value for the enterprise's production needs, but also could be reused efficiently *a posteriori* for other needs. However, it has been shown that hardly 10% of this capital is exploited. This can be explained by the difficulty in reusing information that has been structured for objectives different from capitalization and reuse.

However, big efforts are currently being made to validate these layers of information accumulated as part of data processing for production. The objective is to produce, from these layers, new information that is useful for action within the enterprise, in other words producing "operational knowledge" in the KM sense. This is manipulation of information with the objective of knowledge discovery, called knowledge extraction from data (or "knowledge discovery from data" (KDD)), which is also linked to similar concepts such as "text mining", "data mining" or the "data warehouse".

1.2.1.2.2. Knowledge extraction from texts

Knowledge extraction from texts ("text mining") is the discovery of useful information from "hidden patterns" buried in a large corpus of texts (sometimes called non-structured or semi-structured information). Research engines now process those types of texts, which are more and more abundant and include web pages, a growing source of knowledge.

1.2.1.3. *Supporting technologies for explicit (elicited) KM*

After having extracted part of the knowledge capital, one has an informational corpus that can be used for transferring or operating knowledge. To put this corpus into effect, two kinds of systems can be used, which are as follows.

1.2.1.3.1. Knowledge servers

These are systems, usually part of a company's intranet, through which one can browse (in the most intelligent and ergonomic manner possible) for the elicited knowledge. These systems do not solve problems directly for the users, but give a rich and flexible method for retrieving knowledge that may be useful to solve an operational problem.

1.2.1.3.2. Knowledge based systems

These are computer-based systems that "operate" the elicited knowledge, such as, for example, expert systems. They use the elicited data and structures to solve a precise "high-level" problem: decision support, process supervision, diagnostic, resource planning, design support, etc.

1.2.2. *Tacit KM*

Knowledge elicitation is an approach that may be not chosen by some organizations, for various reasons: difficulties in setting up such processes, which may be long and time consuming, too high a direct cost, confidentiality problems, problems with people, problems with the knowledge networks, etc.

Another possible approach can be derived from the way in which knowledge is produced in organizations, more precisely from the different forms of groups and functions that participate: networks, communities etc., in which knowledge is seen as the result of a cooperative process in a collective action. The problem is then not to elicit this knowledge, but to foster its creation, its sharing by managing the cooperative work of a community of people. One then does not manage the knowledge itself, but the community that creates it. This knowledge may then remain tacit within the community, while being shared and operational. One may then talk of "cooperative knowledge management" ([Soulier 02]).

The tacit/explicit approaches are not opposed but complementary. It is, for instance, useful for a knowledge community that manages its own knowledge to produces visible and tangible records, while, on the other hand, an elicited corpus of knowledge needs a knowledge community to operate it and make it evolve.

Cooperative knowledge management has four key requirements ([Soulier 02]):

– identification of knowledge communities;

– exchange mechanisms that allow knowledge transfer within knowledge communities;

– principles of managing and supervising cooperation;

– technologies supporting cooperative knowledge management.

Most of the time, there are already communities in the company (for specific jobs, in specific professions, for projects, for given practices, specific interests, etc.). One must analyze them and include them in a strategic schema, in order to optimize their performance in terms of knowledge sharing and creation. The problem is to map the communities, either existing or to be developed, so as to define for each community, the issue, the type of the community, the nature of its need or the problem it has to solve.

1.2.2.1. *The functioning modes of knowledge communities*

There are three types of way of working and exchanging information in a community:

– The classic "division of labor". A worker carries out a limited number of tasks, with reduced autonomy in designing and performing the job. This is the most common approach, even in new and modern forms arising from the introduction of information technology. Dedicated structures analyze, design and codify the work to be done. This type of work is supported by formalized links (planning of activities, performance supervision, etc.), which do not facilitate cooperation. Cooperation links are hidden, implicit and limited.

– The "coordination" type ("soft" cooperation). This relies on a better coordination of activities, with information sharing and synchronization of tasks. It is compatible with the Taylor point of view. The only problem is how to do it better and more quickly. The consequence is the implementation of a global information system with a fast communication function and generalized access to information (the archetype is the company intranet and e-mail).

– The "cooperation" type ("strong" cooperation). This involves synergy, not just coordination, between separated tasks. Working together establishes a community with sound bases for:

- the nature of problems and of knowledge that are to be developed;

- the identification of objectives;
- convergence of reasons why people work together.

1.2.2.2. *Knowledge community supervision*

Implementation of knowledge communities may be performed through their life cycle. This leads to the definition of:

– management commitment in the different phases in the cycle;

– indicators for supervision and control;

– adequate supporting tools.

The different phases in a community development are as follows:

– *Emergence:* a community is created based on a knowledge domain. This creation arises from aggregation according to affinities. It is spontaneous (a community cannot be prescribed). Management is not a player, but may be an observer. The main need is connection between players.

– *Structuring:* the emergent community starts to collaborate. It is consolidated by peer co-option. Information sharing leads to coordination, but every player is autonomous. The community is still informal, driven by the participants themselves. Management is still an observer. The needs are for coordination and for mutual information.

– *Becoming official:* the community is structured and active enough to become visible to the organization (if it is too premature, it stops). The community becomes coherent through a shared common goal. Management becomes involved so as to give a sense of the company strategic objectives and to give official recognition, together with resources. The needs are then for collaborative work.

– *Consolidation:* the community works and produces on a regular basis, and this work is integrated in collective action. Management periodically evaluates the actions and productions of the community. The needs are then for sharing and capitalization.

– *Dissociation:* well anchored communities tend to withdraw into themselves, and some outside signs illustrate this (development of a jargon, group thinking). Management must clearly observe these phenomena and must facilitate cooperation and transfer of knowledge between communities. It is necessary to know how to stop a community, in the same way that one can stop a project. This dissociation maybe generated internally by players having their creative capacity blocked and finding in the emergence of another community the necessary rebirth of the expression of their knowledge need.

To pilot cooperation it is necessary to define:

– *indicators of cooperation:* how to recognize that there is a development of cooperation within the different types of collective work (team, project, process and network);

– *indicators of results of the cooperation:* what are the effects of the cooperation on the knowledge capital?

Several criteria seem useful in recognizing that members of a network are cooperating:

– they construct, from their individual representations, shared representations of problems to solve, objectives to reach and goals to achieve;

– they communicate efficiently, using a common language, understanding the language of the others and sharing their point of view;

– they move out of their domain (discipline, sectors, geographical, etc.) in interdisciplinary actions;

– they accept the existence of conflicts (of criteria, of points of view,. of priorities), managing them in an appropriate time and proceeding to arbitration;

– they put some new applicable and evolutionary organization into place.

1.2.2.3. *Supporting technologies for knowledge communities*

The new types of work described above are tightly linked to rapid adoption and dissemination of systems and technologies for information and communication (ICT) within the organization. The rapid increase and development of new ICT has had a considerable influence on knowledge management. There is sometimes confusion (perhaps deliberate) between ICT and KM.

In classical information systems (databases, data banks, etc.), functionalities are well known: an information system is by definition a system for processing, storing and presenting information. Coming from computer networks, communication technology had the clear functionality of data transmission. With the introduction of new types of tools, especially web-based tools, these distinctions are no longer clear; they are more complex and difficult to understand in depth from the user's point of view. In fact, the new ICT products must be analyzed from the point of view of services, and those services justify their use in knowledge management problems, especially cooperative work, in the sense defined above.

There are four types of services related to ICT ([Germain 98], [Bitouzet 97]):

– communication;

– coordination;

– resource sharing;

– information sharing.

Among those services, one can distinguish between the basic services limited to communication software (communication and coordination) and services with added value based on the information system of the company. However, this distinction becomes fuzzy in the new usages of ICT (e-mail may be used as an information base and information bases may be linked to communication tools, as on a website).

This set of services helps the collaborators in organizing knowledge sharing and working in a cooperative way, especially through knowledge communities. However, one must not forget that although a technical platform may be necessary, it is neither a prerequisite nor a goal. Cooperative KM is really a problem of organization and method.

1.3. The key factors of success in KM processes

A KM process is put in place to optimize knowledge use and evolution in an enterprise. It concerns in a crucial manner knowledge players: experts, specialists, competent people without whom the knowledge capital would not have added value. As it is an element of strategic management, management must be strongly implicated at many hierarchical levels. Here resides both the strength and the weakness in knowledge management: cohabitation of two fundamental networks, the knowledge network and the power network. This is why a KM project does not look like a classic project, because two poles of decision, completely different, must cohabit to construct a common system: a KM system.

The setting up of a KM system in an enterprise is therefore delicate, and a certain number of unusual factors must be taken into account. We enumerate some of them here.

1.3.1. *The water lily strategy*

One key factor in a KM project is the project deployment strategy. There is a radical difference compared with a classic project.

Let us take the example of an Information System project. The project deployment strategy is composed of a needs analysis, a specification and a roadmap that predicts a progressive and linear implementation, integrating users progressively through, in most cases, a training program. One now knows the advantages and inconveniences of this kind of project deployment: the more mistakes are upstream in the process, the more costly they are; risks of failure near the end of the process are far from being negligible, early endings are not exceptional, etc. Project management is centralized (in general by a dedicated information systems), the project is linear (for example, the cost of realization of the second half is half the total professional cost), it is planned, users are involved in a sporadic manner (sometimes at the beginning of project, and at the end for training), etc. These types of "roadmap strategies" for a knowledge management strategy (unless, of course, it is reduced to an information system project) are often revealed to be a failure. They are too linear and do not take enough account of incentive factors. A "socio-technical" type of strategy is necessary.

One can build a new type of project strategy, radically opposite to the "roadmap" one. It can be called the "water lily strategy". This term expresses the idea of "steady organic growth", designating the cell growing process (meiosis in morphogenesis), which processes by successive cellular division to create some viable and complex structure.

The water lily evokes a well known childish riddle: "a water lily doubles its surface every day; if it takes 40 days to cover half of the pond, how many days are needed to cover the entire pond?". An imprudent and too quick answer would be "80 days", assuming that as much time is necessary to cover that second half as the first half. This answer neglects the nature of the water lily and the initial hypothesis! If one transcribes this riddle into project management terms, it may become: "a KM project is expensive; if the project costs €40 M to cover the needs for managing half of the Knowledge Capital, how much will be needed for managing the entire Knowledge Capital?". A linear extrapolation of costs will undoubtedly tend toward €80 M rather than €41 M! It is not usual to invoke the hypothesis of the water lily in this kind of problem.

It is, however, the "water lily" strategy that is the most appropriate for a KM project. It is the best way to include the change process, incentive factors, and the emergence phenomenon in complex systems. The "water lily" strategy is a strategy of *constant effort and cumulative effect*, which perfectly suits the cumulative economic nature of knowledge ([Foray 00]). It indicates specifically that if efforts are considerable to start the project (from a human or other resources points of view), they will not be necessarily multiplied during the spread of the project. As incentive factors are essential, one benefits from a leadership effect. The material costs are not necessarily the most important; they do not therefore grow linearly, etc. One of the main requirements of this type of project management is to minimize costs and risks. The first "water lilies" are not too expensive. Stops, often unavoidable on medium-term projects, are not damaging. Failures do not necessarily imply completely revising the entire approach.

The "water lily" strategy usually comprises three phases.

1.3.1.1. *The pilot project(s)*

The first "water lilies" are persuasive projects. They are autonomous projects, concerning a restricted part of the knowledge capital. It is often possible to identify such projects that have already emerged spontaneously (often a long time ago) in the enterprise: it is indeed true that the knowledge acquisition has been (fortunately!) a common and pragmatic practice for a long time. These pilot projects, concerned with knowledge capitalization, sharing or creation can be revealed or initialized by knowledge cartography. Some of their features are as follows:

– They concern a small knowledge domain ("too small"), but meaningful.

– They are performed by people convinced of the necessity to act on their knowledge and who are therefore susceptible to take a leadership role in a general approach.

– They are based and achieved locally in units. They are not necessarily supported by the management (it is necessary, however, to get an "understanding neutrality" on the part of management). This means that the considerable effort involved in getting, from the beginning, "top management commitment", which is often a prerequisite for strategic projects but sometimes a delaying or even blocking factor, can be avoided.

– They show real life examples of the ways in which knowledge capital can be shared.

Pilot project(s) are the yeast of a global and strategic KM approach. They are a significant "bottom up" approach, in which people who produce and use knowledge in their daily practice ("knowledge workers") prove the need and the possibility of capitalizing, sharing or evolving this knowledge within their context. A pilot project must show that working on knowledge is useful, and that original added value is created compared with a more classic project (quality, documentation, data processing, etc.). It provides a concrete and pragmatic starting point for communicating to management and the other people in the company and contributing to awareness in this domain.

1.3.1.2. *The federation of KM projects*

The first water lilies simply emerge, which is a typical phenomenon related to the complexity of a knowledge system. This emergence phenomenon must be carefully managed in the first phase. Then, a life cycle similar to the one described in section 2.2 is followed. Coordination tasks must be undertaken to gather and to federate the different projects, and, while the strategic and global dimension is not still major, it begins to appear. Cooperation tasks then take place, giving an official status to this federation of projects organized as a network (the KM network of the enterprise).

Several points are addressed:

– the definition of a common objective that gives sense to the set of projects, and a unified shared view; this makes it possible to become persuasive and take a leading role for the rest of the enterprise;

– the official status legitimating through actions of communication and explanation by the hierarchies concerned; this makes it possible to give a formal structure to the network, and means to support it, as well as strong recognition of the project;

– regular and coherent communication on KM inside (and maybe outside) the enterprise.

1.3.1.3. *Project deployment*

The network of "water lilies" being now in place, one must ensure its "steady organic growth" in order to cover the set of needs of the enterprise for the management of its knowledge capital.

This project deployment may be usefully initialized by critical knowledge cartography (see below). The set of critical domains and corresponding actions can thus be defined and included in a global plan. KM processes (see below) must be put in place, with assessment tools. Global supervision tools must also be put in place (as balanced scorecards or the Intellectual Capital Navigator). Communication and incentives must be especially studied, because a global project must mobilize a large number of people to enrich and to bring alive the KM system. This last point often implies a deep change in habits and beliefs. Incentives are not always of the classical kind and often rely more on intellectual or non-material satisfactions than on material rewards. The key factors of success are also (and maybe especially) a part of these incentives, which are the essential drivers for change in this type of project. This is the topic of the following paragraph.

1.3.2. *Change factors*

The setting up of a KM system goes through a certain number of delicate phases that must take account of the bivalent nature of the project. These phases are key factors of success, to which it is necessary to pay careful attention. Among these factors are the following:

– *The "mirror step"*: a KM system is built with knowledge holders. One of the first factors of incentive and acceptance for the system is that these people recognize themselves in the implemented knowledge. It must be a structured and valorizing picture of them. This step is essential. If knowledge holders do not recognize themselves in the system, it will not have any chance of being validated thereafter.

This first step allows the concerned “knowledge network” to become a supporter of the project.

– *Consensus*: knowledge implemented in the system must be consensual, reflecting the individual knowledge that is part of it. Consensus is not natural, especially in a company. It requires a specific and sensitive process.

– *Legitimating*: consensus between participants is not sufficient to validate the knowledge put into the system. This cannot be validated in the classic sense, because it deals with personal and collective knowledge that is very particular to the enterprise, at the highest level, and for which an external reference model never exists. The only possible validation is obtaining a consensus from a “peer group”, in general constituted by other players in the knowledge network or related players.

– *Approval*: the knowledge, once legitimized, must be endorsed by the hierarchy that gives it an official and productive status in the organization. This stage is the starting point for project dissemination and for use in the company.

– *Appropriation*: it is clear that a knowledge management system does not have an interest if there is no appropriation and no evolution of the knowledge in the enterprise. This is a final step, which initializes a “virtuous knowledge cycle”. Up to now, very few know how to perform this phase correctly.

This set of key factors shows, if a demonstration is still needed, that KM in an enterprise is a complex process, which must be implemented progressively and with a rather long perspective. In this domain, as in many others, there are no miracle tools or methods capable of solving all problems. In this domain, more than elsewhere, we deal with knowledge, a matter that has concerned human beings since the beginning of mankind!

1.4. KM: an action for continuous progress

KM is not a new concept, fortunately. Right from its start, any enterprise manages its knowledge, its know-how, generating documents and procedures, and disseminating them, for example, via training and organizing exchanges of any form with their collaborators. What is new is the strategic dimension of knowledge, as resource of competitiveness and performance. It obliges the enterprise to have a global, conscious and reasoned approach to its knowledge capital. This is a long-term process to be achieved progressively, starting from the set of actions of the KM type that have already been carried out in the organization (most often without knowing that this is KM). It is also a cultural process that must gradually develop in daily work, and not a revolution that must change everything. In a word, it is a process involving continuous progress that relies on what already exists. This process takes place in several steps that we will outline here.

1.4.1. *Knowledge cartography*

One of the first difficulties in a KM process is the impossibility of correctly visualizing the available knowledge in the enterprise. The adage "one doesn't know what one knows" is truer for organizations than for people.

Upstream of all KM operations, knowledge cartography aims at putting in evidence the knowledge capital of an enterprise. Organizations wanting to manage their knowledge capital must therefore make a careful analysis of this knowledge capital, in order to determine, in their strategy, what are the pieces of knowledge they must preserve, develop, abandon, etc. Cartography then becomes a tool for decision support and it is necessary to elaborate the specific criteria that make it possible to assess, in the cartography, what are the most critical areas of knowledge for the enterprise and why. One then speaks of critical knowledge cartography ([Peil 01]).

The knowledge map is a hierarchical visualization of domains that are useful to professionals in the enterprise. A knowledge domain can be defined as the field of activity involving a set of people for which one can group information and knowledge. The cartography is articulated around axes defining the strategic knowledge domains, often corresponding to missions of the organization. It thus classifies, in a hierarchical manner, domains that form a decomposition of these axes. Knowledge domains can (must) be linked to a set of people who maintain knowledge and information (structured or semi-structured). The model may be enriched by models of actors or information (documentary references, "best practices", etc.). This is especially important, for example, for the realization of a knowledge server. The map is often represented as a graphical model.

The knowledge map built in the previous phase defines the distribution of different knowledge domains according to the missions of the organization. The next phase consists of establishing the critical character of the knowledge domains. The criticality of a domain is an assessment of the risks/opportunities for the enterprise linked to that domain. For instance, there may exist risks of losing knowledge or know-how that could have serious consequences or there may exist interests in developing a domain to obtain advantages for the enterprise (gain of productivity, of market sectors, etc.). One must therefore define what is, objectively, the criticality of a knowledge domain and provide a method of assessment that allows the most critical knowledge domains in the cartography to be marked out.

The criticality is evaluated according to criteria that are not necessarily easy to define. Criticality criteria for determining the strategic importance of a knowledge corpus can be of various kinds, and very dependent on the culture and the situation of the enterprise. Generally, one can distinguish two types of criteria:

– factual criteria for evaluating the nature of knowledge independently of the content of this knowledge (depth, width, complexity, etc.);

– strategic criteria for evaluating the adequacy of the considered knowledge to the missions or strategic objectives of the organization.

1.4.2. *The repository of KM processes*

The knowledge cartography makes it possible to visualize the available knowledge capital on one hand and, on the other hand, via the criticality analysis, to discover, while justifying them, opportunities and threats that are associated with different knowledge domains. It enables a focus, therefore, on the most critical domains and puts priorities on the different aspects of a coherent plan of action for KM.

It remains, then, to put devices in place (organizations, methods, tools) in order to reduce the criticality of the domains concerned. Rather than considering separate devices, which would look more like a panoply of tools than an integrated process, an alternative approach would consist of piloting the action by setting up a KM process, in a classical way, as for quality. Up to now, a consolidated approach to what a KM process can be does not exist, as such an approach exists, for example, for industrial manufacturing or design. KM has still to build a repository of processes whose strategic objective is to reduce the knowledge criticality, according to the type of identified criticality, and whose tactical objective is to bring added value to knowledge capital.

A possible framework is given, for instance, by the "Daisy Model", which defines and details the key processes in KM.

These processes are both internal, such as capitalization and sharing or creativeness and learning, and external, such as environmental scanning or business intelligence, which must start from internal knowledge and feed it back, or customer relationship marketing, which acts as a filter on the immense potentialities of creation and evolution of knowledge in the enterprise.

KM is the management of these processes, and especially the management of their link to the knowledge capital of the enterprise. One can describe them in terms of four generic classes, which correspond to the "petals" of the daisy, and a class that corresponds to the heart of this model (Figure 1.1):

– *The capitalization and sharing process.* This achieves the "virtuous knowledge cycle" and ensures the dissemination (the "recycling") of the knowledge resource in the enterprise.

– *The interaction process with the environment.* A system isolated from its environment is a dead system. This is especially true for knowledge, fed by more and more important information flows, which come from the environment of the enterprise. The process transforms these fluxes of information into knowledge, useful to the enterprise. They include, for instance, the processes of environmental scanning, economic or strategic intelligence ("business intelligence"). Up to now, this kind of process has been based on external information processing, and not on interaction with the knowledge of the enterprise.

– *The learning and creation process.* This is an endogenous and collective process that is the basis of knowledge evolution. It includes the issues concerning the learning organization and of creativeness.

– *The selection process by the environment.* This is an evolutionary process, selecting created knowledge, according to criteria of the market, of acceptability, etc., that is both economic and socio-technical. It includes the issues concerning marketing, customer relationships, etc. The problem of KM is to integrate this type of issue in a strong relationship with the critical knowledge of the enterprise, notably professional knowledge, for example.

– To this set of processes, one can add a fifth that is entirely internal to the knowledge capital. It is an assessment that is qualitative, quantitative, financial, etc. There already exist methods for valuing an intangible, or intellectual, capital. However, there does not exist, as far as we know, a complete process for valorizing, in the broad sense of that term, knowledge capital.

Ongoing research will allow the definition of a set of methods and of coherent tools to manage these processes and to reach the KM objectives (see [Ermine 03]).

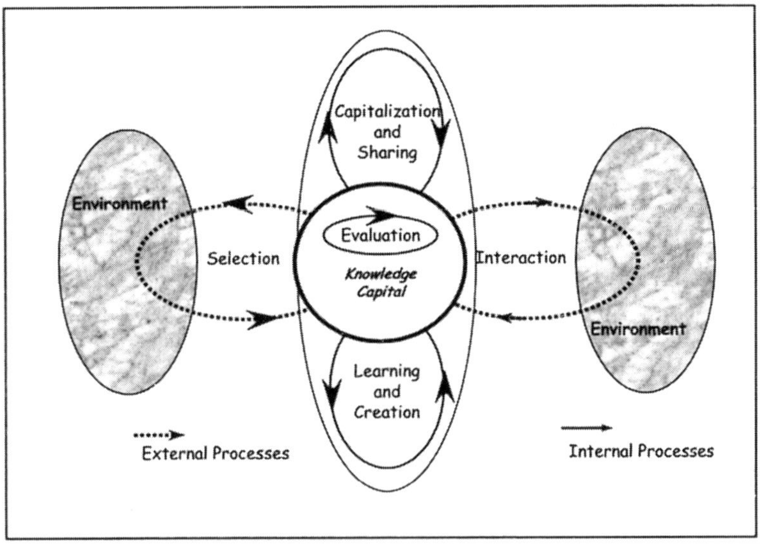

Figure 1.1. *The Daisy Model: key KM processes*

1.4.3. *The KM actions*

After the cartography of knowledge, i.e. the identification of appropriate processes to put in place, has been established, the real operational phase begins. To be efficient and sustainable, it must be integrated into the continuity and process of progress. Enterprises did not wait for the "KM concept – or buzz word" before they began to manage their knowledge. For a long time, they have set up procedures for capitalization, for scanning, for training, for creativeness, etc. The KM processes model, through the Daisy Model, makes it possible to foster and supervise those actions.

Let us take an example: the capitalization and sharing process, which is the most well known and most studied process in KM (Nonaka's process), is a virtuous knowledge cycle (see for example [Ermine 02]), in which a community elicits a tacit knowledge that it possesses; this elicited knowledge is then disseminated in the enterprise and, at the end, this elicited and shared knowledge is appropriated by actors in the enterprise who recreate their own operational and personal know-how. Therefore, there are three subprocesses: elicitation, sharing and appropriation. There is also an alternative to this process of capitalization and indirect sharing and this is a direct process in which the tacit knowledge of actors is shared directly with other actors, who create for themselves their own tacit operational knowledge. This alternative process may be supported, for example, by face-to-face training, communities of practice, shared experience, etc.

For a given knowledge domain, where the criticality is caused by a bad capitalization and/or bad sharing, one can analyze the problem via either the direct or the indirect process. For example, for the indirect process, the questions are: what the procedures have been put in place for elicitation? For sharing? What procedure guarantees that actors appropriate knowledge that is available from the knowledge bases in their daily work? And so on. These simple questions permit the identification of hiatuses, the definition of some reasonable progress axes on the basis of those that already exist, and the identification of solutions that should be put in place in this process.

These solutions are extremely numerous and must be analyzed deeply before a decision is made, because the corresponding investments are often significant. Let us give some examples of possible tools:

– for the indirect capitalization and sharing process;

– for elicitation:

- knowledge modeling (knowledge books, knowledge engineering, story telling, etc.),

- knowledge transcription:

- secondary documents,

- authoring: cards (return on experience, quality documents, etc.), documents, reports, publications, etc.),

- etc.,

– for sharing:

- collaborative spaces,

- digital diffusion (push, pull, profiles, etc.),

- document diffusion,

- browsers,

- etc.,

– for appropriation:

- operational training,

- e-learning,

- etc.,

– for the direct capitalization and sharing process:

- knowledge communities,

- seminars, meetings, convention,
- expert 'Yellow Pages',
- etc.

1.4.4. *Piloting processes and actions*

A global and strategic implementation of KM in the enterprise can be piloted from elements that have been described here: critical knowledge cartography, KM processes, axes of progression, procedures and actions.

One then needs to set up a supervision process, including the surveillance of processes with adequate indicators and a global assessment. This assessment, in order to be consistent with a KM policy, must show the value added that is brought to the knowledge capital of the enterprise, and the progression of the enterprise in its "knowledge maturity".

As tools of assessment, one can give as an example the grids of maturity elaborated by the French KM club ([Club 00]).

1.5. Conclusion

KM is not a new concept in enterprises. Since social organizations have existed (that is to say since the beginning of mankind!), human beings have always tried to maintain a collective memory, to share some useful knowledge, and to constantly create new knowledge. It is the cumulative and strategic dimensions of knowledge identified in some new problems that have revealed the emergence of new fundamental factors.

Among these new problems, some have had a negative outcome, described in illustrative stories or more personal anecdotes, often painfully experienced. They concern episodes in which the knowledge capital, which has been discussed above, has been squandered thoughtlessly. It can be seen in reorganization plans, in massive layoffs, in waves of departure for retirement, in business process reengineering, in badly managed mergers or acquisitions. Badly transmitted knowledge which is badly shared and badly preserved has already brought much disillusion and loss, as much at the productivity level as at the human level.

There have been other more positive developments, for instance the major evolution of means of information processing and communication. Information technology has switched from storing, representing and retrieving data to a situation that enables information sharing, resources sharing, coordination and communication.

We have moved from an era where information was a very rare and strategic commodity to an era where it is overabundant and widely available, where the sending and receiving structures have become more important than messages themselves. This evolution has brought some considerable gains, and triggered a wild race in the implementation of these new technologies, which must now, after the first euphoria, search for a new framework within the classical information paradigm.

The "KM" paradigm emerges as an option, because what is now important is not obtaining information, but ensuring that this information is operational in a particular context and at a given moment, for the right person who must solve a complex task with his or her necessarily limited cognitive capacity. This is what is meant by the expression "the right information, at the right moment, to the right people". And this only describes half of the problem, which is the use of knowledge as productive resource; the other half is, of course, the production and the evolution of knowledge, without which one cannot speak of the use of knowledge capital.

KM is therefore a new way of management. It is the management of an intangible asset, which is a strategic asset of the organization, via flows of new type (cognitive flows, which must be differentiated from information flows) that are either generated by actors in the organization (competency flows) or used by these actors (cognition flows) in their production activities, these actors becoming real "knowledge workers".

1.6. Bibliography

[Bitouzet 97] Bitouzet C., Fournier P., Tézenas du Montecel B., *Management et intranet*, Hermès, 1997.

[Club 00]: Club Gestion des Connaissances, Ermine J.-L., Laude H., *A Knowledge Maturity Model*, Proceedings of "Knowledge Management; Theory and Practice" Workshop (J.-L. Ermine ed.), in PKDD'2000 (Principles of Knowledge Discovery from Data), 12 September 2000, Lyon, pp. 13-18.

[Dieng 00] Dieng R., Corby O., Giboin A., Goliebowska J., Matta N., Ribière M., *Méthodes et outils pour la gestion des connaissances*, Dunod, Paris, 2000.

[Edvinson 99] Edvinson L., Malone M., *Le capital immatériel de l'entreprise*, Ed. Maxima, Paris 1999.

[Ermine 02] Ermine J.-L., *Initiation à la méthode MASK*, CD-ROM from Troyes University of Technology, France, 2002.

[Ermine 03] Ermine J.-L., *La gestion des connaissances*, Hermes Science Publications, 2003.

[Foray 00] Foray D., *L'économie de la connaissance*, Repères, no. 298, Ed. La découverte, Paris, 2000.

[Germain 98] Germain M., *L'intranet*, Economica, 1998.

[OCDE 96] OCDE, *Mesurer le capital humain, vers une comptabilité du savoir acquis,* 1996.

[OCDE 99] OCDE, *L'économie fondée sur le savoir, des faits et des chiffres*, 1999.

[Peil 01] Peil O., Aubertin G., Ermine J.-L., Matta N., *La cartographie des connaissances critiques*, un outil de gestion stratégique des connaissances. CITE'2001, Coopération, Innovation et Technologies, 29-30 November 2001, Troyes, pp. 287-308.

[Soulier 02] Soulier E., Zacklad M., Ermine J.-L., *La gestion coopérative des connaissances,* EGC'2002, Extraction et Gestion des Connaissances, Actes de la session industrielle, pp. 59-72, 2002.

[Stewart 97] Stewart T., *Intellectual Capital, the New Wealth of Organizations*, Currency/Doubleday, New York, 1997.

[Strassmann 98] Strassmann Paul A., *The Value of Knowledge Capital*, American Programmer, March 1998.

Chapter 2

Can One Identify and Measure the Intangible Capital of the Enterprise?

The world of economics is changing in depth. Leverages of development are moving from matter and energy to information, knowledge and organization, in other words, from tangible to intangible. The rapid growth of information services, which are invading the industrial sphere more and more, are driving a new way of articulating offer and demand, one that is co-productive in nature and which appears to be at the heart of processes of value creation. The recent explosion, without precedent in our history, of costs related to intangible investment, that is, dedicated to R&D, training, software, market exploration and organization, has put design and knowledge management activities in a central position in the performance of enterprises as well as nations. The multiplication and the extension of networks, while ensuring the storage and the diffusion of information on a large scale and at an ever-decreasing cost, has the tendency to transform the world into an immense informational spider's web. This has the effect of displacing efficiency towards network structures and relational systems. We are entering, at least in the industrialized countries, a transition phase, from an economy founded on rationalization of material activities and intensive energy resources to an economy founded on information and knowledge, the stakes in which are considerable for enterprises.

These enterprises face an environment that is more and more uncertain. Since the 1950s, their environment has changed drastically. The globalization of economics and of markets and the growing importance of networks are factors that have

---

Chapter written by Patrick EPINGARD.

contributed extensively to the destabilization of economies and to the rise of uncertainties. The steady and repetitive characteristic status of information during that period has gradually been replaced by an erratic flux of "perturbations" that prevent economic agents from predicting, even approximately, the evolution of variables in their environments. This is most often a pure, radical uncertainty that is not computable using statistics.

This strong unpredictability is a consequence of the increasing complexity of economic systems (from the Latin word *complexus*, meaning woven together): the multiplication of products and services, of real or potential actors, of relationships with partners or competitors (which are often entangled), as well as the increasing complexity of technologies that have to be mastered. It generates a strong uncertainty about the future, even the near future, and requires a necessary mobilization of the cognitive resources of individuals and especially of organizations. Facing this increasing complexity requires the production and the management of knowledge, itself more and more complex, created from the mobilization and the cohesion of a set of individual expertises [De Bandt 2001]. This is precisely the objective of Knowledge Management (KM).

But we need to go beyond the acceptance, which is today extensively shared, that the intangible capital of the enterprise has become a key factor of performance, or that information and knowledge constitute some strategic resources for organizations. We propose in this chapter to address the problem of measuring and assessing the intangible capital of the firm. It is, as we will see, an asset of a very different nature from physical capital. Can one measure this asset? Can one value its global or partial (through its components) impact on performance? Can one distinguish the effects of physical capital and intangible capital, or must the measure be global? In other words, what can we presently measure, and what do we want to measure?

Before discussing some ways to answer to these difficult questions of measure and assessment, we need to identify adequate concepts, in order to avoid building on quicksand. What is the intangible capital of an enterprise? What are its components? What makes the difference from physical capital? Is it sound to consider a knowledge "stockpile" in the enterprise? What are the components of this "stockpile"? Is it necessary to distinguish information and knowledge and, if yes, what is the theoretical and practical range of this distinction? These questions require a clarification, to avoid regrettable confusions, which are too frequent in the present literature, as we can see. To clarify these concepts we will use recent theoretical advances that analyze the firm as a processor of knowledge in the so-called "knowledge-based economy".

2.1. The intangible capital: an essential and elusive concept

We need to refer to the basic theoretical sources. The economic notion of "capital" [Cota 1991] sends us back to the idea of a set of possessions serving to produce other possessions. We are driven very quickly to distinguish between the capital circulating and the stationary capital, the latter being composed of lasting indirect possessions that do not disappear physically during the process of production (machines, buildings, infrastructures, etc.).

To deal with the relative imprecision of the extension of that concept (must we, for example, include human capital, financial capital, etc.?), we need to make a clear distinction between capital and production and to specify their relationship. Capital is a stock that has a given value at instant t, while production is a flux (or income) that has a duration, from time t to time t + 1. The production of equipment goods, which is related to investment, is therefore a flux that feeds the capital. The relationship is as follows: the capital at time t is enriched in period to t + 1 by the gross investment in the period, but is also reduced as a result of economic depreciation:

$$K_{t+1} = K_t(1-a) + I_{(t;t+1)}$$

$$\Delta K_{(t+1;t)} = I_{(t;t+1)} - aK_t$$

where a is a capital depreciation factor and $I_{(t;t+1)}$ is the gross investment during the period (t;t+1). We van see, therefore, that if, for example, the depreciation of the asset is greater than the gross investment during the period, the value of the capital decreases. We can also see that if, after having defined the concept of capital, we need its measure, i.e. to value the capital, it is necessary to measure both the gross investment during the period and the depreciation of the capital. To do that, we assume that we know how to value the capital at the instant t, which does not seem obvious for intangible capital.

How can we distinguish physical capital and intangible capital? Cotta (1991) proposes defining the goods of the intangible capital as "the set of knowledge that allowed the creation of the existing material and intangible goods and that allows today the use of material goods". These intangible goods would be different from material goods, being a set of goods that are either financially or physically created by people in their effort to coping with their environment. Material or intangible, these goods can be "long" or "short", depending on whether their longevity is greater or less than the duration of the production process (we recognize here the distinction between circulating capital and fixed capital).

This definition emphasizes the notion of "knowledge", as in the definition of intangible investment I have already proposed (in [Epingard 1999]): joint production by knowledge creation that is incorporated in a sustainable manner in people (human capital), technical objects (technological capital) and organization (organizational capital). If one makes the analogy with physical goods, the intangible investment would be therefore a "flux" of knowledge that feeds the intangible capital that one can assimilate into a "stock" of knowledge.

Is it really as simple as this? We do not believe that it is, because this analogy relies on a supposed isomorphism between merchandise and knowledge, which we have so far accepted without analyzing it, the founding cumulative hypothesis of the positive theory of capital implying that some knowledge is added to existing knowledge, other knowledge disappears and the value of intangible capital is an arithmetic consequence. This hypothesis seems false to us, because knowledge does not reduce to merchandise.

At this stage, it is important to distinguish carefully between information and knowledge, terms that are too often considered, in a very reducing way, as interchangeable. According to our concept, information is clearly located in the domain of what is possible to codify. It takes the form of a set of formatted and structured data that (and we must take account of this from now on) cannot generate new information. It is the raw material of knowledge, the flux of messages that increases it, restructures it or modifies it. However, as K. Boulding had already noticed more than 30 years ago, knowledge must not be considered as a simple "stock", in the sense of a container that is filled and emptied from fluxes of information, but rather as a "structure" that feeds himself with information, that may become richer by its use.

Knowledge is a cognitive capacity and a learning capacity. It requires the exploitation of information, pre-existing cognitive categories, interpretation codes and various faculties (judgment, intuition, ability, etc.). What is rare today is not information, which is available at lower and lower cost (it is the cost of creation that is important, the marginal cost of reproduction is nearly zero), but the capacity and expertise to sort it out and to exploit it, which relies on cognitive capacities. Knowledge is therefore far from being reducible to information, as supposed implicitly by the standard theory, which is extensively based on the technical metaphor of information as a signal that it is merely emitted and transmitted to a receiver, at the lowest possible cost and without distortion. Moreover, a frequent mistake in the literature consists of considering that knowledge increases in linear proportion to the accumulated information, which implies that the dimension of interpretation is completely ignored. There is a big difference, in particular from the point of view of cost, between duplication of information (marginal cost nearly zero) and transfer of knowledge (cost often high). Knowledge is not, in most cases, separable from its human support and is often difficult to transmit beyond its initial production context, especially because it is partly tacit.

From this major distinction, and if we come back to the definition of intangible capital as set of knowledge, we can wonder, on the one hand, what is the nature of knowledge in the enterprise and, on the other hand, where it is stocked.

So, what is an intangible good? The most usual definition in the literature is a definition by negation: an intangible good or asset is a non-material good, which means non-observable or, more precisely, one of which only the effects can be observed, like the invisible man whom one only knows by traces left on cushions (when he removes his bandages, of course!) [Demotes-Mainard 2003].

Is it satisfactory to say that goods are material and services are intangible, wonders J. De Bandt (2002). This would force us to recognize implicitly that the expression "intangible goods" is an oxymoron, a contradiction in terms. However, one must recognize that all goods which possess a physical envelope by definition incorporate, in their value chain, intangible components to a more or less important degree and, at present, to a more and more important degree. For example, the physical support of a book or a software program has very little value compared to the content, i.e. to the activities of design and production of the embedded knowledge. Then one has to decide where to place the cursor; to what degree of intangibility, in the "goods–services continuum", can one speak of "intangible goods"?[1]

The "intangible capital" of the enterprise is for us a set of assets that can be described in terms of expertise and knowledge that may be able to produce a flux of incomes for a long time. We then have to determine, on the one hand, how to locate such a set of assets within the enterprise and, on the other hand, how to build a typology of intangible capital that makes sense for cost measurement and value creation in the enterprise.

For knowledge location, Machlup (1984) distinguishes three kinds of storage of knowledge, corresponding to three kinds of capital:

– knowledge integrated in tools and machines;

---

1 P. Dumesnil [1995] writes: "[the] present use of the expression 'capital' or 'intangible investment' appears to us perfectly paradoxical. It designates, most often, the embedding of the language in something 'material'. That this material is concerned with micro-physics and can be embedded and reproduced at low cost, while using 'small' quantities of energy, makes it by no means intangible." This is why we consider that the qualifier "intangible" does not describe the support, but the content in terms of knowledge of the productive asset that is implied.

– knowledge retained by individuals;

– knowledge that circulates because it is not restricted to a specific knowledge holder; it can be freely disseminated, or not, such as, for example, scientific knowledge.

It seems to us that Machlup disregards an essential support of knowledge storage:[2] the organization. This term must be understood in two ways: as a means of coordination within a set of rules on the one hand and as a collective entity on the other hand. Organizational knowledge is distributed among actors and is, at the same time, crystallized in structures. It can in no way be reduced to the sum of individuals, even though they may be very rich, and even though, of course, the knowledge of the organization is built from the knowledge of the individuals that compose it. It includes a set of coordination and decision rules, explicit or tacit (routines, habits), as well as the collective know-how embedded in the organization. It also includes a "collective memory" supported by the information system of the enterprise. This is why organizational learning is now an important topic: an organization can learn and store its knowledge in a collective memory. This way of learning and developing fundamental expertise is a strong leverage for performance.

To spotlight organizational capital, the KM pioneering authors Edvinson and Malone proposed the equation:

intangible capital = human capital + structural capital

the latter being defined by the authors as "What remains in the enterprise when employees have gone home" (client files, licenses, information systems, quality, etc.). A major difference between human capital and structural capital is, of course, that the first does not legally belong to the enterprise, while the second may be an object of negotiation, which is essential from the creditors' point of view. It should be noted that although structural capital is essential for competitiveness, it is almost completely absent in the private accounting of enterprises (one cannot presently activate expenses that are permitted to constitute it). If one wants to measure it and to value its effects on competitiveness, one needs an extra type of accountant, which is an unsolved problem.

Theories that are more especially interested in the effects of intangible capital on economic growth, and in particular theories of endogenous growth, whose leading idea is that technical progress depends on the choice of agents so that a more or less

2 We emphasize the locations of knowledge retained by the enterprise. However, only a part of the "knowledge stockpile" is localized, while another part is distributed, as a result of the knowledge diffusion process.

important part of resources can be dedicated to the costs of R&D or training are especially interested in two components of it, linked to two types of specific externalities:

– *Human capital*, which includes the set of knowledge and ability (expertise) embedded in individuals, having the potential to increase their efficiency, must be distinguished from physical capital on at least two points: it is not appropriable by others (property of exclusive rights and non-rivalry), and it is far less limited than physical capital, from the point of view of innovation and creativeness, as well as learning (it is produced from itself). We should also note that human capital is a much more opaque entity than physical capital, even though quality signals (diplomas, title, etc.) may limit this opacity, which generates assessment problems.

The source of enrichment of human capital is obviously in education and learning, but authors such as Lucas (1988) also integrate in their model a hypothesis of knowledge externality, the idea being that the capital human of an individual is much more rich and efficient if he or she can share and exchange that knowledge with other people whose human capital is itself rich. There is, in fact, a positive externality of networking for human capital: while increasing his or her level of expertise, the individual also increases that of others, via the enriched quality of their exchanges, and this is reciprocal. This also happens at an inter-businesses level, as for instance in Silicon Valley. The problem is that, even though one is conscious of the importance of these knowledge externalities, it is difficult to measure them, and in particular to isolate their effect on growth rate.

– *Technological capital*, which is initiated by R&D investments, is considered by Romer (1990) as a semi-public good, which means that it is in one part appropriable (protection and profit making by the sale of patents) and in another part non-rival (the possession of knowledge by an individual does not prevent another individual also having it). It is precisely this character of non-rivalry of technological knowledge that makes it possible to increase scalable yields: not being incorporated intrinsically in a physical object, knowledge can be used in numerous activities at a very weak marginal cost (compared to the cost of creation). When, for example, one constructs two identical factories, one only incurs the cost of the innovation once.

Technological externality is related to R&D, and more extensively to the innovation in process or product. It is based on the fact that every agent can take advantage of the available technological knowledge, knowledge that is itself the result of a long process of intangible capital accumulation over time. This technological externality is both interagent and intertemporal.

The effects of technological capital and human capital are combined to increase the productivity of the enterprise. The knowledge they carry, which is both the input and the output of technical progress [Encaoua *et al.* 2004], can permit some increasing yields because it possesses attributes that are fundamentally different from a traditional good. Let us recall its two main distinctive features.

It is a public good[3]: being essentially non-rival, it needs to be produced only once, but it is not limited in use. One does not lose acquired knowledge and hence that "input" has a multiplicative character. There is an important point to be made here: because of its specific features and, in particular, the weakness of its marginal cost of use, knowledge can be incorporated simultaneously into several supports. Thus, the knowledge produced during the R&D phases is included in technical objects, but it can also be kept by the enterprise in tacit or codified form. On the one hand, there is no simple link between an intangible investment type and an incorporation type, while, on the other hand, it would not make sense to want to "add" the human capital to the technological capital of the enterprise so that its value appreciates, since the substance is in part common. We will come back to this point.

It is cumulative: we saw that R&D generates positive externalities, because present research is extensively based on past research. We noted that information cannot generate any new information, whereas knowledge has this property of being able to generate new knowledge and to become richer when it is used. Research also improves learning capacities and existing knowledge absorption, which finally becomes an intangible asset of the enterprise. One sees the importance of the cumulative process that treats knowledge as input and as output of the production process.

The intangible capital in the enterprise is constituted of a set of knowledge and ability of both human and technological nature, plus the quality of the organization (broadly speaking), which economic models hardly integrate because of its non-measurability. However, it plays an essential role in the performance of enterprises.

This corporate knowledge is partly codified and partly tacit, especially the know-how. An important part of the codified knowledge comes from tacit knowledge conversion, which means that it is not expressible outside the action of the person who retains it, because it is embedded in its human support[4]. Through knowledge

3 This does not mean that some knowledge bases are not private, but that essentially knowledge is non-rival, and that consequently it is not possible to guarantee an efficient knowledge production process based exclusively on a market system, since sale to the marginal cost (near zero or very weak) would not permit the producer to be remunerated. See Foray [2000].

4 The importance of this tacit knowledge and the difficulty in codifying it can be seen in the process of creation of an expert system. The expert is not indeed expert on its own expertise and a part of the knowledge that he or she retains, like his or her ability, is not easily expressible. To codify this knowledge is therefore an extremely complex operation, involving

codification, humanity exploits this unique faculty that he or she possesses, according to A. Leroi-Gourhan, "to place his memory outside of himself". But the expansion of codified databases, due to the emergence of ICT, does not suppress the existence and even the necessity of tacit knowledge. One could even say that it reinforces them, because codified knowledge development causes learning activities (needed for instance to exploit the knowledge) and new knowledge emergence and ability, which reinforce and so on. Consequently, tacit and codified knowledge, far from neutralizing one another, are in constant conversion[5], and are involved in a dynamic interaction relationship that, via organizational learning, can initiate virtuous spirals and increase the value of the intangible capital of the enterprise.

Intangible investments, such as R&D, training, software creation and marketing, contribute to the constitution of assets that do not necessarily have a commercial value outside the organization, which means that they are specific assets in Williamson's sense. In fact, the more the constituted asset is specific and the more tacit the embedded knowledge is, the farther we are from a notion of merchandise exchangeable on a market, and vice versa[6]. If one takes out a license or a patent, this is one example of generic knowledge that is extensively codified, for which there exists a market. On the other hand, the casual networks or a team's collective ability are very specific assets and the knowledge that they generate has a predominantly tacit part. These specific assets, which are fundamental for the performance of the enterprise, i.e., cannot be evaluated on the market. Between those two extremes (the specific and perfectly identifiable asset on the one hand and the specific and non-valuable – cannot be valued – asset on the other), there exist a multiplicity of intermediate intangible assets characterized by the human dimension of their incorporation (knowledge partly codified and partly tacit) and by the constituted asset specificity (information system of the enterprise, for example, often partly generic and partly specific). These intangible assets are in general at least partially transferable on the market, and therefore are partially valuable, i.e. they can be valued. This explains the difference between the accounting value of assets and the global acquisition value of an enterprise; we will examine this point below.

The field of the intangible is, therefore, particularly difficult to define and to evaluate, partly because of the hidden part of the iceberg, the most important part probably, which is not directly measurable. However, the problem is that enterprises

not also translation, but also re-creation supported by a dedicated language. This is why tacit knowledge and codified knowledge will never totally overlap.

5 See the pioneering work of Nonaka [1994], who proposes qualifying the transformation processes for knowledge as follows: externalization for transformation from tacit to explicit and internalization for explicit to tacit.

6 For a classification of intangible assets, see Duizabo and Guillaume (1996).

feel the need to understand their key factors of success, of which they strongly recognize the intangible character, and to evaluate the contribution to their performances. The problem is that intangible asset measurement remains at present a real theoretical enigma.

2.2. Immaterial measurement: a theoretical enigma?

What do we know and what are we likely to want to measure? What we can observe currently, when looking at the existing statistics on this question, is more a mosaic of measures concerning partial elements, which are not always coherent, rather than a global approach: "What appears to be missing is a global approach that would integrate the different steps in a coherent way, to build a real observation system", Mr Demotes-Mainard (2003, p. 1), a French national specialist (INSEE) in statistics about the information society, has pertinently commented.

However, these scattered and heterogeneous statistical analyses are not the only obstacles to an appropriate understanding of the phenomenon. Difficulties in measuring intangible assets are intrinsically related to the specific features that differentiate them from physical assets, and essentially to the three following features:

– *Non-separability*: it is often very difficult to isolate assets, or even to separate the material and intangible aspects and to value them separately. Under these conditions we must consider the global value; analysis carried out element by element is meaningless. The problem is that even if we adopt a global approach, the specificity of the intangible asset usually prevents the establishment of an organized and efficient market that would permit the assignment of a price, in contrast to financial or material assets. Assessment by the market of intangible asset is therefore nearly impossible [Alcouffe 2003].

– *The undetermined life cycle of these assets*: it is much easier to observe the material wear in a machine than the loss of value due to wear or obsolescence of human capital. Furthermore, human capital has the property that it becomes more valuable when used, via learning, which is what principally differentiates it from physical capital. This does not mean that the risk of obsolescence is negligible with regard to intangible assets; in a world of uncertainty and high rate of change, we see a kind of "Schumpeterian" process of creative destruction, of unceasing creation of new elements and destruction of old elements that constantly feeds back accumulated knowledge and expertise. Maintaining intangible capital thus becomes essential; it relies on the management of competences and know-how in organizations and, therefore, of course, on knowledge management.

– *The opacity of these assets, even if only partial*: intangible assets are most often very specific and are embedded in their human support, so that their quality is not directly observable; this generates a large uncertainty concerning future income, which remains even when a market valuation exists. Let us think, for example, about the market in life-long learning and certain types of service that have observable costs (often high, for example, training related to human resource development). Their impact on the constitution of sustainable intangible assets, and therefore on the performance of the enterprise, remains very uncertain.

Even if we consider the recent evolution of accounting practice, which is aimed at partially integrating, under some restraining conditions, the expenses of R&D, software, brands and intellectual creation, we are far from obtaining a faithful accounting of the value of the intangible assets of the enterprise. In fact, accountants, guided by the prudent principle of only accounting for elements that are part of the legal capital of the enterprise, are led by legal, financial and economic logic, which necessarily creates a bias in their judgment on what really constitutes the economic assets of the enterprise, and they can only provide a distorted picture.

From a strictly accounting point of view, intangible capital is characterized, not only by the so-called "incorporeal" assets (patents, licenses) that are the visible part of the iceberg, but by the goodwill or overvalue (economic value – patrimonial value) that characterizes the potential that the enterprise has created over time (a set of scientific, technological and commercial expertise, as well as quality of management, social climate, know-how, etc.). Formally, if B is the benefit of the enterprise, CNA the corrected net asset and i the rate of sure investment of capital, the goodwill is equal to k(B – i.CNA), where k is a rate that allows this "super-profit" to be capitalized over several years. Here we see the idea, developed by the Austrian economists that the value of the enterprise depends on its future potential (goodwill) and not on its past states (net assets valued to their historic value).

The overvalue therefore represents the gap between the price paid for the acquisition of an enterprise and the accounting capital value of the enterprise. It is thus assumed to represent a measure of the value of the intangible assets purchased at the date of acquisition. From a financial point of view concerning the goodwill, one can consider the intangible asset as a fund, the value of which depends on its capacity to generate profits during its life cycle. If, for instance, the enterprise reduces the accounting value of its goodwill, previously registered as part of its assets, this means that it is noting the loss of value of its intangible assets. It is therefore clear that, in order for the corresponding investments to be worth considering, the value of constituted asset must be greater than its patrimonial value.

However, a problem remains: outside accounting conventions, what do we measure and what do we know we are measuring? When we speak of the value of an asset, it is very important to distinguish the input, which means what is performed, and the output, the result of the action. In other words, it is necessary to distinguish the expenses of intangible investment and the value of the capital that is constituted as a result of this investment. There are some completely unproductive intangible investments, for example, expensive training, that do not have any effect, direct or indirect, on the profitability of the enterprise. Conversely, one may sometimes have inexpensive investments (a slight reorganization, for example) that make possible a very appreciable improvement in the performance of the enterprise. Let us add that quality of strategic management and leaders' choices have a determining effect. This explains why accounting only for expenses related to intangible investments is never sufficient to estimate their value.

If inputs are considered, they can be partially measured using a system of statistical observation. Thus, intangible investment by French enterprises has been measured in France since 1974. Initiated by the Crédit National and then followed by Insee, these measures have been based on the centralized data produced by the statistical services (Ministry of Research, Ministry of Industry, etc.) and address four components: R&D, software, continuous training and commercial investment. Without going into the detail of the statistical analysis, the main conclusions of this survey, which are also valid for other industrialized countries (particularly the USA and the countries of northern Europe), are the following:

– From a modest initial level, one can observe, at the macro-economic level, a growth rate of the intangible investment (measured from the aggregation of the four components above) significantly greater than that of the physical investment (measured by the overall value of the fixed assets). Thus, the ratio of intangible investment to GFFC increased from 21.1% in 1974 to 39.1% in 1986 and 57.2% in 1996. One can see, therefore, that according to these measures just in the end of the 20th century, the expenses of intangible investment represented about a third of the total investment in enterprises (material + intangible).

– The ratio discussed above tends to stabilize since, according to the most recent measures carried out by the Ministry on the industry (that have been reduced since 1996, which makes the historic sequence somewhat heterogeneous), it would have been 47.5% in 1998 and 41.2% in 2000. We are not therefore in presence of a phenomenon that would lead progressively to the disappearance of the physical investment in enterprises. We are rather in the presence of a complementarity phenomenon: the intangible relies on the material (there is no "information-based" economy without a "material-based" economy) and transforms it by feedback, as the informational density of the physical capital becomes stronger and stronger.

– Finally, we can see a very strong sectorial concentration of intangible investment, in particular for R&D and learning. Since 1995, more than 60% of the total cost of R&D has been in five sectors (electronics, aeronautics, automotive, chemistry and pharmacy). In these sectors the rate of investment in R&D is nearly equal to the rate of physical investment. With regard to training, the effort of enterprises has also varied: since 1994 the percentage of the total salaries dedicated to professional development ranged from 1.88% (agriculture, forestry, fishing) to 12.2% (air and spatial transportation), with a strong concentration of the high rates in hi-tech enterprises, which may lead us to conclude that there is a strong relationship between the two types of investment (technological and training). This concentration of the relationship is of course linked to the evolution of the size of the enterprise: the 50 largest French groups represent more than 60% of the cost of R&D. Enterprises with 2,000 salaried employees or more (30% of salaried employees) represented 52% of the total costs for continuous training in 1994. These rates can be observed in all industrialized countries.

However, if the very strong growth of intangible investments in the last 30 years is real, it is necessary to validate what is measured and what is the actual capacity of juxtaposing scattered elements, so that the economic value of the intangible investment is faithfully reflected. It is clear that the measure is only about part of the range, that which is *a priori* relevant to measure. This is why the investment in strategic analysis and the organizational investments, whose impact is often primordial, are not included in these measures. We should note that organization, because it determines the extent of knowledge diffusion in the enterprise, is located at the centre of the innovation process, an interactive process that requires intense communication between the actors, as well as much feedback between domains such as science, manufacturing and marketing.

Furthermore, beyond these partial and necessarily unsatisfactory measures of intangible input, how can we measure the value of the output, i.e. the resultant productive intangible asset? We know that this value depends strongly on how these different components are composed, on their organization and their consistency. But here is the weakness of these additive measures: they do not address the organizational dimension. From that point of view, economists and accountants have the same difficulties, as we can see with the difficulty in valuing impacts on the economic growth of technological capital or human capital, because it is impossible to assess their synergies.

Depending on their context and specific point of view, an actor will adopt specific approaches for assessing the intangible. We will distinguish and consider here the point of views of three typical characters: the creditor, the manager and the economist.

The creditor is interested in security, usually when a legal right is established and when an asset is separable. Legal protection can apply only to separable intangible assets as a patent or royalty. It is not applicable to intangible assets that are not separately transferable, for example, a team's specific know-how within the organization. Between the two poles, there exist some intermediate situations concerning partially separable and partially protected goods (software integrated in a machine, for example), but it is clear that the creditor's very legalistic point of view is oriented and partial, since it is unaware of the intangible assets that are essential for economic efficiency, because they are not the legal property of the enterprise (Demotes-Mainard 2003, p. 6).

The manager, who has many difficulties in making measurements because he or she lacks encoded reliable and transparent information on the intangible assets of enterprises [Bonfour 2000], essentially needs to value inputs, i.e. the necessary resources to constitute the technological, human, organizational capacities and, more globally, the "reputation capital" of the enterprise (its brands). One can therefore try to construct indicators that allow the evolution of these different types of expense to be followed, through reporting practices. As we have seen, managers can hardly rely on financial accounting that does not really consider elements that contribute to the constitution of the intangible assets of the enterprise.

This is why one sees developing, in enterprises anxious to make a better valuation of the contribution of intangible capital to the creation of value, a set of practices centered on the dynamic management of the intangible capital. The approach used involves constructing indicators[7], monetary or other, that make it possible to follow the evolution of the intellectual capital in the enterprise, or its most important components: human capital, "memory and methods" capital, customer relationship capital, development capital, etc. These indicators are used for management and may be a basis for the definition of strategies based on intangible resources. They are the foundation of the method initiated by Skandia and extensively used and adapted by numerous enterprises through the world.

As Bounfour [2000] says, a number of managers must become aware of the necessity to change a static vision of the organization as "stock of resources" to a dynamic vision, founded on an evolutionary approach, of the organization as a "dynamic combination of capacities". This means that in a vision placing intangible capital at the centre of the approach, one must pay attention above all to processes of value creation, which are closely integrated to the strategic approach. This is in one way a return from the performance-related approach of procedural rationality to an approach that follows the sense of H-A Simon: the quality of the process of problem solving is more important than results achieved in the short term, because it determines the competitiveness of the enterprise in the long term. Such an approach

7 For concrete examples of indicators by type of capacity (human, technological, etc.) and the way to measure them and to link them to the performance of the enterprise, see Bounfour [2000].

emphasizes the cognitive aspects of organizations and encourages paying careful attention to individual as well as collective learning processes.

For such a purpose, it is necessary to adapt management tools and instruments, which is not easy. Management instrumentation is today in a deep crisis. The massive irruption of these new realities, as the rise of intangibles is deeply transforming the productive systems, imposes in-depth changes on the way of managing the enterprise, and especially on the ratio of action efficiency and conditions of its measurability, as a result of questioning paradigms that underlie the use of traditional management tools.

Let us take the concrete example of investment choices. Is one able mechanically to transpose the traditional assessment methods of the profitability of the investment to the intangible? Is one seriously able to calculate the internal rate of return (IRR), the net present value (NPV) or the delay in ROI of a training action or an investment in R&D, which are most often located very upstream of the manufacturing process and the market sale? Let us recall that the possibility of valuing these criteria of project selection in a meaningful way assumes that the following conditions are fulfilled.

The parameters of choice, such as cost and induced effects, must be valued in monetary terms, as a set of expenses and gradual benefits. To be pertinent, it is necessary that this reduction to the monetary aspects of the costs and advantages does not let the essential aspects, which are qualitative in such projects, escape.

It must be possible to value these same parameters in differential terms with respect to a reference situation (usually no action). This requires an approach that involves a given and stable organization, which is very problematic in periods of rapid change.

It must be possible to connect effects to the causes that generated them, which assumes the establishment of a direct causality link between initial expense and induced cash flows. This link is particularly loose and indirect when intenagibles are concerned, since, as we have emphasized, the impact of intangible investment is assessed in terms of creation of potential. If this is not the case, this approach assumes that investments are analytically separable, which is incompatible with an integrated system logic, the performance of which can only be assessed globally.

There is thus a universe of unknown variables faced by those who try to measure the profitability of an investment with intangibles playing a major part. Such an investment is very different from the simple substitution of an old machine by a new one. Benefits ascribed to such investments are only partially the result of direct output. Above all, there is an improvement of quality, a shortening of development and manufacturing cycles, a better coordination of functions, the networking of operations and devices, the learning of new techniques and new kinds of organization (Lorino 1991) that play an important role in performance. Effects produced by this type of investment, projects with a strong integrated organizational dimension, are essentially indirect and long term.

Basing the calculation on a fixed and stable organization, like traditional calculations of profitability, would be unsuccessful, because it misses the main aspect: there are organizational changes as well as learning activities in complex projects. Some authors (Favereau 1989) then make attempts to build an "organizational economic evaluation" that would assess, with suitable indicators (monetary or not, and even qualitative for the strategic elements that it is impossible to quantify), the global impact of the project on the organization, with a horizon of assessment that allows the immediate financial effects to be by-passed, in order to improve the quality of the investment decision in a framework of a widened rationality. The construction of a methodology more specifically oriented toward the intangible within the framework of a procedural rationality remains, however, for the greater part a task that has still to be undertaken, even though some proposals are beginning to appear in research reviews on management.

Economists who work in the framework of a knowledge-based economy try, first of all, using a macro-economic approach, to measure the impact of the intangible capital on growth and productivity. We have already quoted theories of the endogenous growth (P. Romer, A. Lucas, etc.) that are alternatives to the traditional theories of growth, of neoclassical obedience (R. Solow), because the former assign a secondary role, i.e. an exogenous role, to research and education,

Without detailing these models (see Epingard 1988), we will mention that they start from the principle that growth can, in the framework of a knowledge-based economy, maintain itself thanks to the existence of increasing yields of knowledge. Even though this hypothesis is not always validated empirically, it has a strong impact and is based on the specific features of knowledge as input to the production function. Not being incorporated intrinsically in a physical object, and being a public good by nature, as we analyzed above, it possesses a demultiplying property that can be reused to produce very weak marginal cost on several supports. Certainly, if A are the non-rival inputs (the knowledge) and X the rival inputs (machines, men) of the production function, one can suppose that:

$$F(A, \lambda X) = \lambda F(A, X).$$

According to the economist B. Arthur (1996), these increasing yields of knowledge essentially concern the hi-tech world (pharmaceutical products, telecommunications, information technology, aeronautics, etc.) characterized by products that are very demanding of "knowledge resources" and demand very different strategies from products requiring "material resources". In the latter case, traditional cost–volume optimization is required, while in the former, anticipatory strategy, connections to the cognitive networks and exploitation of positive feedback linked to the increasing yield of knowledge of others are necessary.

Without going into detail about the diverse impacts on the productivity of the intangible capital, let us indicate that they converge to show the existence of the following:

– *A strong link between R&D and productivity*. The rate of output of R&D would be equal to, or in number of cases even significantly greater than the rate of output due to the physical capital. The weakness of this measure remains, however, the rhythm of depreciation of the capital. According to the chosen hypothesis of obsolescence of knowledge, the rate of output may vary from "very high" to "low" and it is very difficult to know what is most realistic among these hypotheses. Also note that measures of the technological and R&D externalities show that they are important and that, in particular, the social output of research is, in most cases, extensively superior to its private output, which reveals a technological knowledge diffusion process that is both inter-enterprise and inter-sector.

– *A strong link between human capital and productivity*. We know that specificity of human capital explains, to a large extent, the heterogeneity in the performances of enterprises. But we also know that it is much more difficult to measure the productivity of the human capital than to measure that of the technological capital, because, as we have seen, knowledge incorporated in people is to a large part tacit and specific. So, we are brought back to the problem of measuring knowledge inputs and outputs, and the difficulties involved in constructing the *ad hoc* indicators are huge: how to measure fluxes and stocks, moreover including aspects of non-codified knowledge, how to appreciate the quality of knowledge diffusion in the economy, how to measure the obsolescence of knowledge and expertise. As far as human capital is concerned, it is clear that the traditional indicators (years of study, qualification) are insufficient, since they very reflect badly expertise acquired by the individual through a learning process. Up to now, we are not aware of any attempt at quantification of the meta-investment, strategic in a knowledge-based economy, which constitutes the learning itself [Cover, 1996].

2.3. Conclusion

In our attempt to define the concept of "intangible capital", we can note its strategic importance both for the economist and for the manager, and the major difficulty encountered in delimiting the concept, because of its specificities in relation to physical capital. We are in the presence of an intangible object, knowledge, which possesses some very particular properties, and the functioning and the effects of which we can directly observe. Furthermore, all would be simpler if the set of the knowledge accumulated by the enterprise existed in a codified form, but we know that the enterprise is essentially a learning organization that includes to a high degree an irreducible tacit dimension. Moreover, because its form is its essence, knowledge does not possess additivity properties that would allow it to be assimilated into the physical capital, so it must be managed separately, in a way that is adapted to its specificities: "knowledge is organized according to a specific structure, specific functions and cohesion", says KM specialist J.-L. Ermine (1988). This requires specific strategic plans for knowledge management that are aimed at creation, capitalization and sharing of the intellectual capital of the enterprise.

In our question on the possibility of measuring intangible investment and intangible assets of the enterprise, we also face major difficulties. Accounting measures are too reducing and too partial, while the instrumentation of management in its whole, which is designed to value economies founded on the rationalization of material activity, ignores qualitative dimensions such as the quality of the staff, the accumulated knowledge, the relationship networks, both internal and external, the strategy, etc. Using the language of the economist, we know how to measure a part of the inputs of the function of knowledge production (some cost), but the output, and especially the precise link between input and output, is still mainly unknown.

However, even if there is still strong uncertainty about problems of measurement and evaluation of the performance of intangible asset, it is certain that knowledge capital is a key factor in the success of enterprises as well as of nations. This awareness has driven numerous enterprises to develop sophisticated knowledge management tools, even if the development and the exploitation of these tools is itself a very expensive process in terms of cognitive resource consumption. This proves that the challenge is considered to be sufficiently important to justify this intangible investment.

Today, new theoretical approaches are emerging, alternatives to the standard theory, with the aim of renewing concepts and methods so that the intangible dimension of our economies can be better addressed. In the theory of the firm, for example, the approach that considers the firm not as a machine to maximize a functional objective in all circumstances (the so-called "black box" approach), but as a processor of knowledge, relies on the principle that the firm follows a behavior function of accumulated expertise during its history. Centered on the learning processes, knowledge creation and knowledge sharing processes, this approach appears fully compatible with KM, and one can predict with no risk that it will develop in the coming years.

2.4. References

Alcouffe C., Louzzani Y. (2003), "Mesures de l'investissement immatériel: indicateurs d'évaluation et de performance", note no. 381, LIRHE.

Arthur W.-B. (1996), "Increasing Returns and the New World Business", *Harvard Business Review*, July-August, pp. 100–109.

Bounfour A. (2000), "La valeur dynamique du capital immatériel", *Revue Française de gestion*, September–October, pp. 94–110.

Carter A.-P. (1996), "Measuring the Performance of a Knowledge-Based Economy", in OCDE, *Employment and Growth in a Knowledge-based Economy*, Paris.

Cohendet P., Llerena P. [1990], "Nature de l'information, évaluation et organisation de l'entreprise", *Revue d'économie industrielle*, no. 51, pp. 145–165.

Cotta A. (1991), "Le Capital", in *Encyclopédia Universalis*, pp. 929–932.

De Bandt J. (2001), "De l'information à la production de connaissances et de valeurs", in J. De Bandt and G. Gourdet (Eds), *Immatériel, nouveaux concepts,* Economica.

De Bandt J. (2001), "Réflexion sur l'immatériel", in J. De Bandt and G. Gourdet (Eds.), *Immatériel, nouveaux concepts,* Economica, Paris, p. 24.

De Bandt J., Gourdet. G (Eds) (2001), *Immatériel, nouveaux concepts*, Economica, Paris.

Demotes Marnard M. (2003), "La connaissance statistique de l'immatériel", Colloque de Voorburg sur la statistique des services, Tokyo, 6–10 October.

Dumesnil P. (1995), "Economie de la langue et langage de l'économie", *Revue internationale de systémique*, vol. 9.

Duizabo, R., Guillaume N. (1996), "Approche d'une nouvelle typologie des actifs immatériels", *Echanges*, no. 119, January, pp. 34–39.

Edvinson L., Malone M. (1999), *Le capital immatériel de l'entreprise*, Maxima, Paris.

Encaoua. D., Foray D., Hatchuel A., Mairesse J. (2004), "Les enjeux économiques de l'innovation", *Revue d'économie politique*, no. 114(2), March–April.

Epingard P. (1998), "Etude d'un objet conceptuel déstabilisant: l'investissement immaterial", *Revue économique*, no. 49(6), pp.1511–1547.

Epingard P. (1999), *L'investissement immatériel, cœur d'une économie fondée sur le savoir*, CNRS-éditions, Paris.

Ermine J.-L. (1998), "Capter et créer le capital savoir", Réalités industrielles, *Annales de Mines*, pp. 82–86.

Favereau O. (1989), "Vers un calcul économique organisationnel?", *Revue d'économie politique*, 99(2), pp. 322–335.

Foray D. [2000], "L'économie de la connaissance", *Repères* no. 302, La Découverte, Paris.

Guellec D., Ralle P. (1995), "Les nouvelles théories de la croissance", Repères no. 161, La Découverte, Paris.

Lorino P. (1991), *L'économiste et le manageur*, La Découverte, Paris.

Lucas R. (1988), "On the mechanics of economical development", *Journal of Monetary Economics*, vol. 22, pp. 3–42.

Machlup F. (1984), "Knowledge: its Creation, Distribution and Economic Significance", *The Economics of Information and Human Capital*, vol. 3, Princeton University Press.

Ministère de l'Industrie (2003), *Les chiffres clés de l'industrie française*, Paris.

Nonaka I. (1994), "A Dynamic Theory of Organizational Knowledge Creation", *Organization Science*, vol. 5.

Romer P. (1990), "Endogenous Technological Change", *Journal of Political Economy*, vol. 98, pp. S71–S102.

Simon H.-A. (1976), "From Substantive to Procedural Rationality", in S.J. Latsis (Ed), *Method and Appraisal in Economics*, Cambridge University Press, pp. 129–148.

Chapter 3

Complexity Theory: Dynamics and Non-Linearity are the Only Reason for Knowledge Management to Exist

3.1. Introduction

A lot has been said and written about knowledge management, probably starting with the proponents of the learning organization on the one hand, and Nonaka's view on Knowledge Management (KM) on the other hand. Increasingly, authors have added the subject to their vocabulary and the more that the "general management thinkers" have become involved (Leonard-Barton, Drucker, etc.), the more KM has acquired the status of a major buzzword. At the 1999 European Conference on Information Systems (Copenhagen) the "best research paper award" was given to a paper that argued that KM would be the next hype to render people obsolete (Swan *et al.*; 1999). This choice appeared to me to represent a public act of masochism on behalf of the IS community, given that IS experts, more than any other people, should have a clear idea of why KM is here to stay.

This chapter attempts to provide a broad framework for the subject, highlighting the different aspects (including the human ones) that should be considered when talking about KM. This "taxonomy in brief" is of course based on a particular paradigm (like any other taxonomy) that is known as the complexity paradigm. Looking through the lenses of complexity theory, we can see why

---

Chapter written by Walter BAETS.

KM is a new and fundamental corporate activity. Complexity theory allows us to understand why knowledge is a corporate asset and why and how it should be managed. The lenses of complexity theory allow us to say that KM is not just another activity of importance for a company.

A number of KM projects, based on this taxonomy, have been researched over the last five years within Notion (the Nyenrode Institute for KM and Virtual Education), a research center fully sponsored by Achmea (the second largest Dutch insurance holding; the fifth largest within its European network), Atos/Origin, Philips, Sara Lee/DE and Microsoft. Full details of these research projects can be found in Baets (2005).

This chapter attempts to present a complete picture of KM, starting with the paradigm and covering the infrastructure and process, with the aim of clarifying the subject of study. Both the corporate and the academic perspectives appear in this chapter.

3.2. The knowledge era

An important and remarkable evolution in what we still call today the industrial world is that it is no longer industrial. We are witnessing a rapid transition from an industrial society into a knowledge society. The knowledge society is based on the growing importance of knowledge as the so-called fourth production factor. Many products and certainly all services have a high research and development cost, whereas the production cost itself is rather low. Developing and launching a new operating system like Windows costs a huge amount of investment for Microsoft, which makes the first copy very expensive, but any further copies have a very low production cost. Having a number of consultants working for a company is a large investment for a consulting company, so when they are actively working on a project, their marginal cost is close to zero. Having the knowledge base, which means having the consultants available, is expensive. Their real work for a client is relatively cheaper. Even the best example of industrial production in the Western part of the world, which is car manufacturing, has become increasingly knowledge-based. More than 40% of the sales price of a car is attributable to research, development and marketing.

We continue to talk about the industrialized countries, since most of our thinking is still based on concepts of industrial production dating back to the earlier parts of the 20th century (if not even the end of the 19th century). What we have observed, though, is that increasingly companies get involved in optimizing supply chains and that those supply chains evolve into demand and supply chains. The following step consists of supporting those chains with information technology (IT) in order to increase efficiency. The strange thing that happens in a following stage is that a progressive use of IT puts pressure on the very existence of the chain. The better a chain is integrated on the basis of IT, the more a pressure gets created that makes the chain explode into a network. Particularly in such circumstances, the "owner" of the knowledge base manages the process. Network structures evolve around knowledge centers. Companies manage brands and outsource most of the chain itself. Extreme examples of this approach are probably Calvin Klein, Benetton and Nike. Again, knowledge and particularly the capacity to manage, create and share knowledge is becoming the center of the scope of the successful company. This can be translated via brand management, direct marketing to targeted clients, etc., but it is the visible part of the evolution from an industrial market into a knowledge-based market. Knowledge becomes yet another attribute of the changing economic reality.

Knowledge in a company has different forms and most commonly one regroups these forms into three categories of knowledge. Tacit knowledge is mainly based on lived experiences while explicit knowledge refers to the rules and procedures that a company follows. Cultural knowledge then is the environment in which the company and the individual (within the company) operate.

Different forms of knowledge are crafted by various different activities. Conversion of knowledge takes place based on the tacit and explicit knowledge that a person possesses or has access to. The creation of knowledge very often takes place during joint work sessions, such as brainstorms, management meetings, etc. Equally important but more difficult to capture is knowledge processing via assimilation. Very often, assimilation is based on cultural knowledge as a first input, reinforced with tacit knowledge that quite often collapses with explicit rules and regulations. It seems important to stress, however, that KM is only the "sufficient" condition. The "necessary" condition for dealing with new economic realities is the boundary condition for KM and that is the learning culture of the company. On top of the mere fact that the most interesting knowledge is implicit and therefore "stored" in people, it is the dynamics of the knowledge creation and sharing activity (let us for simplicity call this learning) where the people come into the picture for a second time.

Above all, KM and learning is an attitude and a way of working with management. It is an overall approach that goes beyond the addition of a number of functional tactics. One could even say that it is a kind of philosophy of management, rather than a science. This process is one of redefining the target of the company from a profit making or share-value increasing entity to a knowledge-creating and sharing unit. The first type of organization has a rather short-term focus, whereas the latter type has a more visionary and long-term one.

The aim of the company is no longer purely growth as such, but rather it becomes sustainable development and renewal. Hence, organizations not only need knowledge, they also need the skills and competencies to dynamically update and put knowledge into practice. This results in the need for organizations to learn continuously and to look for continuous improvement in their actions through the knowledge acquired. Hence, organizations should embrace the philosophy of the learning organizations, the process being organizational learning (Baets, 1998).

A learning organization enables each of its members to continuously learn and helps to generate new ideas and thinking. By this process, organizations continuously learn from their own and others' experience, adapt and improve their efficiency towards the achievement of their goal. In a way, learning organizations aim to convert themselves into "knowledge-based" organizations by creating, acquiring and transferring knowledge so as to improve their planning and actions.

In order to build a learning organization, or a corporate learning culture, companies should be skilled at systematic problem solving, learning from their own experience, learning from the experiences of others, processing knowledge quickly and efficiently through the organization and experimenting with new approaches. Developments in information and knowledge technologies make it increasingly possible to achieve these competitive needs and skills.

3.3. The complexity paradigm

In the past, and easier to identify when market change moved more slowly, we got used to thinking in terms of reasonably linear behavior as markets and industries appeared to be more stable or mature. Concretely, we thought we could easily forecast future behavior based on past observations and for many applications we developed complex (and sometimes complicated) methods to extrapolate linear trends (Prigogine and Stengers, 1988; Nicolis and Prigogine, 1989). But, in reality, markets do and did not behave in a linear way. The future is not a simple extrapolation of the past. A given action can lead to several possible outcomes ("futures"), some of which are disproportionate in size to the action itself. The "whole" is therefore not equal to the sum of the "parts". This contrasting perspective evolved from complexity and chaos theory. Complexity theory challenges the traditional management assumptions by embracing non-linear and dynamic behavior of systems, and by noting that human activity allows for the possibility of emergent behavior (Maturana and Varela, 1984). Emergence can be defined as the overall system behavior that stems from the interaction of many participants – behavior that cannot be predicted or even "envisioned" from the knowledge of what each component of a system does. Organizations, for example, often experience change processes as emergent behavior. Complexity theory also tells corporate executives that beyond a certain point, increased knowledge of complex, dynamic systems does little to improve the ability to extend the horizon of predictability for those systems. No matter how much one knows about the weather, no matter how powerful the computers, specific long-range predictions are not possible. Knowing is important, not predicting; thus, there is no certainty (Stewart, 1989; Cohen and Stewart, 1994).

The focus on the non-linear behavior of markets contradicts the traditional positivist and Cartesian view of the world. That positivist perspective translated into the traditional management literature – the stuff that most MBAs are taught – describes "the" world in terms of variables and matrices, and within a certain system of coordinates. Exact and objective numbers are needed in order to create models while simulations can offer a "correct" picture of what to expect. Business schools have particularly welcomed this "scientific" way of dealing with management problems as the one which could bring business schools up to the "scientific" level of the beta ("hard") sciences. It is clear that much of existing management practice, theory and "remedies" based on the positivist view is limited by their dependence on several inappropriate assumptions because they do not reflect business and market behavior. Linear and static methods are the ones that are taught in business schools. Therefore, markets have to be linear and static. As we know, they are not (Arthur, 1990).

It seems important to elaborate a little more on positivist thinking as, later, we want to propose a different paradigm.

A major aspect of positivism is the division between object and subject. This means that the outer world (e.g. an industry) is pre-given, ready to be "truthfully" represented by organizations and individuals. The mind is able to create an inner representation that corresponds to the outer world, be it an object, event or state. Translated to knowledge, positivism considers that knowledge exists independently of the human being who uses it, learns it, transfers it. Knowledge reflects and represents "the world in itself" and can be built up independently of the observer, the "knower". What if the universal knowledge that is transferred is mainly a theoretical framework, a form that is of little use in the non-linear and dynamic markets?

Another premise of positivist thinking is based on a strict belief in (absolute) causality and (environmental) determinism. As there are clear-cut connections between cause and effect, managerial actions lead to predictable outcomes and thus to control. Successful systems are driven by negative feedback processes toward predictable states of adaptation to the environment. The dynamics of success are therefore assumed to be a tendency towards stability, regularity and predictability. The classic approach to strategy illustrates this reductionism. The complexities of industries are reduced in terms of maturity, continuity and stability so that a single prediction of an organization's future path can be described. As a consequence, the better the environmental analysis according to a number of dimensions, the better the course (strategy) that can be defined and implemented (Baets and Van der Linden, 2000, 2003).

My own research over the last years, and currently undertaken in the EcKM, suggests that instead of searching for causality, the concept of synchronicity (being together in time), often referred to as a quantum structure, allows much more insight into business dynamics (Baets, 2004). Indeed, that quantum structure is a holistic concept of management, based on interacting "agents". Those networks of agents/people create emergent behavior and knowledge.

Positivism is the prevailing scientific view in the Western world, since it perfectly coincides with the Cartesian view of the world: the over-riding power of man as a fact of nature. Nature gives man the power to master nature, according to the laws of nature. In 1903, however, Poincaré, a French mathematician, cast some doubt in this positivist view. Without really being able to prove it, or even to gather evidence, he warned:

> Sometimes small differences in the initial conditions generate very large differences in the final phenomena. A slight error in the former can produce a tremendous error in the latter. Prediction becomes impossible; we have accidental phenomena.

This suggested that, with the approaches used, humanity was not always able to control its own systems. Hence, there is the limit to the Cartesian view of the world.

It took quite a number of years until, in 1964, Lorenz showed evidence of the phenomenon. Lorenz, an American meteorologist, was interested in weather forecasting. In order to produce forecasts, he built a simple dynamic non-linear model. Though it only consisted of a few equations and a few variables, it showed "strange" behavior. A dynamic model is one where the value in a given period is a function of the value in the previous period. For example, the value of a particular price in a given period is a function of its value in the previous period. Alternatively, the market share for product A in a given period is a function of the market share in the previous period. In other words, most, if not all, economic phenomena are dynamic. Such a dynamic process that continuously changes can only be simulated by a stepwise procedure of very small increments. It is an iterative process. Once the value of the previous period is calculated, it is used as an input value for the next period, etc.

A computer allowed Lorenz to show what could happen with non-linear dynamic systems. As is known, he observed that very small differences in starting values caused chaotic behavior after a number of iterations. The observed difference became larger than the signal itself. Hence, the predictive value of the model became zero (Stewart, 1989). Lorenz's observation caused a real paradigm shift in sciences. Lorenz showed what Poincaré had suggested, namely that non-linear dynamic systems are highly sensitive to initial conditions. Complex adaptive systems are probabilistic rather than deterministic, and factors such as non-linearity can magnify apparently insignificant differences in initial conditions into huge consequences, meaning that the long-term outcomes for complex systems are unknowable. Today we know, thanks to the integration of ideas of the two main scientific revolutions of the last century (relativity and quantum mechanics), that another underlying problem, aggravating the complex structure, is the structure of synchronicity in the "business nature".

Translated to management, this advocates that companies and economies need to be structured to encourage an approach that embraces flux and competition in complex and chaotic contexts rather than a rational one. Mainstream approaches popularized in business texts, however, seldom come to grips with non-linear phenomena. Instead, they tend to model phenomena as if they were linear in order to make them tractable and controllable, and tend to model aggregate behavior as if it is produced by individual entities that all exhibit average behavior.

Positive feedback has been brought into the realm of economics by Brian Arthur (Arthur, 1990), who claimed that there are really two economies, one that functions on the basis of traditional diminishing returns, and one where increasing returns to scale are evident due to positive feedback. Marshall had already introduced the concept of diminishing returns in 1890. This theory was based on industrial production, where one could chose out of many resources and relatively little knowledge was involved in production. Production then seemed to follow the law of diminishing returns, based on negative feedback in the process and this led to a unique (market) equilibrium. Arthur's second economy includes most knowledge industries. In the knowledge economy, companies should focus on adapting and recognizing patterns, and building webs to amplify positive feedback rather than trying to achieve "optimal" performance. A good example is VHS becoming a market standard, without being technically superior. A snowball effect ensued, which made VHS the market standard, even though Betamax offered better technology at a comparable price.

Arthur also specified a number of reasons for increasing returns that particularly fit today's economy. Most products, being highly knowledge intensive, with high up-front costs, network effects, and customer relationships, lead to complex behavior. Let us take the example of Microsoft Windows. The first copy of Windows is quite expensive as a result of the huge research costs. Microsoft experiences a loss on the first generation. The second and following generations cost very little comparatively, but the revenue per product remains the same. Hence, there is a process of increasing returns.

Two more interesting developments have consequences for our argument. Recent neurobiological research, e.g. by Varela (Maturana and Varela, 1984), has revealed the concept of self-organization and the concept that knowledge is not stored, but rather created each time over and again, based on the neural capacity of the brain. Cognition is enacted, which means that cognition only exists in action and interpretation. This concept of enacted cognition goes fundamentally against the prevailing idea that things are outside and the brain is inside the person. The subject can be considered as the special experience of oneself, as a process in terms of truth. By identifying with objects, the individual leaves the opportunity for the objects to "talk". In other words, subject and object meet in interaction, in hybrid structures. Individuals thus become builders of facts in constructing contents of knowledge that relate to events, occurrences and states. Knowledge is concerned with the way one learns to fix the flow of the world in temporal and spatial terms. Consequently, claims of truth are transposed on objects; the subject is "de-subjectivised". There is not such a subdivision between the object and the subject. Cognition is produced by an embodied mind, a mind that is part of a body, sensors and an environment (Baets, 1999; Baets and Van der Linden, 2000).

Research into artificial life gave us the insight that instead of reducing the complex world to simple simulation models, which are never correct, one can equally define some simple rules, which then produce complex behavior (Langton, 1989). This is also a form of self-organization, like the flock of birds that flies south. The first bird is not the leader and does not command the flock. Rather, each bird has a simple rule, e.g. to stay 20 cm away from its two neighbors. This simple rule allows us to simulate the complex behavior of a flock of birds.

Probabilistic non-linear dynamic systems are still considered deterministic. This means that such systems follow rules, even if they are difficult to identify and even if the appearance of the simulated phenomenon suggests complete chaos. The same complex system can produce, at different times, chaotic or orderly behavior. The change between chaos and order cannot be forecast, nor can the moment at which it takes places, either in magnitude or direction. Complexity and chaos refer to the state of a system and not to what we commonly know as complicated, i.e. something that is difficult to do. The latter depends not on the system, but more on the environment and the boundary conditions. Perhaps for a disabled person, driving a car is more complicated. In general, building a house seems more complicated than sewing a suit, but for some other people building a house would be less complicated than sewing a suit. This depends on the boundary conditions for each individual person.

To formalize in a simplified way the findings of complexity theory, we could state three characteristics. First, complex systems are highly dependent on the initial state. A slight change in the starting situation can have dramatic consequences in a later period of time caused by the dynamic and iterative character of the system. Second, one cannot forecast the future based on the past. Based on the principle of the irreversibility of time (of Prigogine), one can only take one step at a time, scanning carefully the new starting position. Third, the scaling factor of a non-linear system causes the appearance of "strange attractors", a local minimum or maximum around which a system seems to stay for a certain period of time in quasi-equilibrium. The number of attractors cannot be forecast, neither can it be forecast when they will attract the phenomenon.

There are a myriad of insights we can gain from complexity theory and its applications in business and markets for KM (Baets, 1998; Baets and Van der Linden, 2000).

The "irreversibility of time" theorem suggests that there is no best solution. There are "best" principles which one can learn, but no best solutions or practices that one can copy. There are even no guaranteed solutions that can be used in most circumstances. This fact deems the need for a different way of organizing the process of knowledge creation and KM.

3.4. What should be understood by KM: the corporate view

Let us recall that this chapter is attempting to present a complete picture of KM, starting from the paradigm and covering the infrastructure and process, with the aim of clarifying the subject of study. Though the corporate and the academic perspectives are at times a little different, they both appear in Figure 3.1.

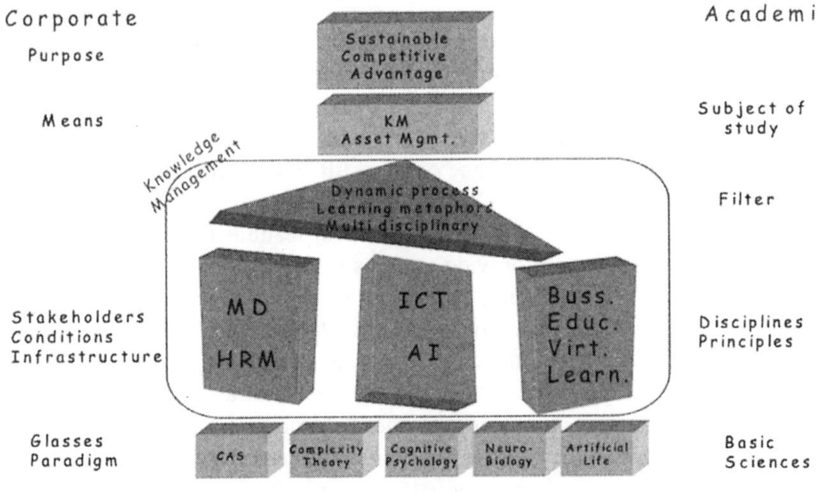

Figure 3.1. *A taxonomy of KM*

Any managerial concept is based on a particular paradigm and, according to the view developed in this chapter, the paradigm of complexity (non-linear and dynamic systems behavior) sheds interesting and refreshing light on the nature of KM. Earlier in this chapter we explained why the complexity paradigm positions knowledge at the center of a knowledge-based company and it does so increasingly with virtual or extended companies.

The left-hand side of Figure 3.1 shows the corporate logic in understanding KM. The paradigm serves as the glasses through which we look at the corporate purpose (gaining sustainable competitive advantage, or expressed more simply, survival) and what we observe then is the means to achieve this aim, i.e. asset management. The chosen glasses allow us to identify the way ahead in reaching the goal. The immediate "next" step is identifying the "infrastructure" or stakeholders necessary for KM:

- human resources management and management development;
- information and communication technology, in particular artificial intelligence;
- business education and (virtual) learning.

The corporate aim remains to create sustainable competitive advantage, and the means for realizing this is (and has always been) asset management. However, for knowledge-intensive companies, this means that KM moves into the picture. A translation (a filter) above and beyond the necessary integration of infrastructure and stakeholders is necessary in order to combine the infrastructure with KM. That filter is a dynamic process, in which the "learner" should be given responsibility. Pedagogical metaphors give us an insight into this filter process (Baets, 1999).

The prevailing pedagogical metaphor is the transfer metaphor. Knowledge in general and, more specifically, subject-related material are viewed as transferable commodities. A student (a learner) is seen as a vessel positioned alongside a loading dock. "Knowledge" is poured into the vessel until it is full. Whereas the student is the empty vessel, the teacher is a crane or a forklift. The teacher delivers and places knowledge into the empty vessel. Courses applying the transfer theory are likely to be very much lecture-based, to include talks from leading figures in the relevant fields (the more the better) and provide students with duplicated course notes. Once the vessel is full, a "bill of loading", which is the diploma, certifies the content of the vessel. IT improves the speed of loading (with high-tech cranes). Nobody can guarantee that in the next port, the cargo is not taken out of the ship. Monitoring a student means monitoring the process of filling the vessel and sometimes sampling the quality of the contents. This same metaphor became the prevailing one in talking about (virtual) KM approaches (Baets and Van der Linden, 2000; 2003).

However, since knowledge appears to be dynamic and learning non-linear (based on our paradigm), another paradigm is necessary. Here again educational science provides us with a valid illustration. The traveling metaphor is one by which the teacher initiates and guides the students through an unknown terrain that needs to be explored. The student is the explorer and the teacher/tutor is the experienced and expert traveling companion and counselor. The guide not only points out the way, but also provides traveling maps and a compass. The "teaching methods" (if one can still call them such) that are most used in applying this theory are experiential methods: simulations, projects, exercises with unpredictable outcomes (as in some case studies), discussions and independent learning. In courses applying this theory, monitoring means regularly comparing each other's traveling notes. Experiments have shown that this theory is particularly effective in adult education, since adults are better equipped in order to deal with the increased responsibility that the "learner" has in this paradigm. One step on from the traveling theory is the growth metaphor. In many respects, this theory does not differ greatly from the previous one. Rather, it is an extension of it and focuses more on the self-initiative of the student. Subject matters are a set of experiences that each student should incorporate into his/her personality, which it is the aim of the student to develop. This latter paradigm (be it the traveling metaphor or the growth one) perfectly fits complexity theory (our overall paradigm or "glasses"). It allows us to integrate the infrastructure into asset management. It introduces the rationale for work-place learning and the necessary integration of the latter with KM. This makes KM different from information management and adds value to it.

3.5. Research perspective on KM

The combination of infrastructure (with its different stakeholders and/or disciplines) and the learning process (filter) makes KM what it should be. Most existing KM theories either do not get much further than a discussion of means and purposes, or they overstress one of the infrastructural aspects, ignoring the unity of and the necessity for all the three elements together. In our view, KM, knowledge creation and knowledge sharing (via virtual learning platforms) are integral parts of the same model.

From a research perspective, we consider complexity theory as the basic science(s) involved. In particular, the following concepts are of importance for the correct understanding of the paradigm and its consequences for KM:

– sensitivity of the complex system to initial conditions;

– existence of (many) strange attractors in complex systems;

– principle of the irreversibility of time (Prigogine);

– behavior of complex systems far away from the equilibrium (Prigogine);

– learning behavior of systems;
– autopoeisis (Varela);
– embodied mind (Varela);
– enacted cognition (Varela);
– research into artificial life and its applications (Langton);
– law of increasing returns (Arthur);
– quantum structure of business.

All these aspects need a good explanation, while a clear link to managerial consequences is necessary.

As already mentioned earlier, and visualized in Figure 3.1, the disciplines involved in KM are human resources management and management development, ICT and particularly artificial intelligence (AI), and business education, increasingly virtual education. The management development (MD) function should be the driver in this knowledge creation, knowledge sharing, learning process, ensuring that each individual receives at the pace that she or he can process. MD should also provide the learning conditions. It is unavoidable that ICT and AI are necessary in order to support the KM process (Baets and Venugopal, 1998; Venugopal and Baets, 1995). Building IT platforms, extracting knowledge via AI and virtual education, are only some of the aspects where IT is of help. Business education, and this increasingly includes virtual education, is responsible for creating some input in the learning process but equally needs to make some of the extracted knowledge accessible to each individual. Business education in this respect has also to do with the content. The aspect of knowledge sharing is an educational one too. KM therefore needs to integrate successfully disciplines such as human resources management, organizational sciences, educational sciences, artificial intelligence and cognitive sciences, thus implicitly defining a KM research agenda.

It is my firm belief that in the decade to come, we will see a breakthrough in the understanding of the underlying theory justifying the (corporate) necessity for KM, in line with the agenda set out in this chapter. As suggested above, the consequence of the concepts developed here and its logical extension is an unavoidable ontological discussion about causality versus synchronicity. In my work (Baets, 2004) I call this the quantum structure of business (or, in particular, in the reference, of innovation), which provides an integrated and applicable theoretical and conceptual framework for understanding and consequently managing dynamic processes, KM only being one of these. The first research projects undertaken confirm this potential understanding and its application in business. It is the acceptance of the ontological evidence for synchronicity that drives the research agenda of EcKM.

3.6. References

Arthur B., Positive Feedbacks in the Economy, *Scientific American*, February, pp. 92–99, 1990.

Baets W., *Organizational Learning and Knowledge Technologies in a Dynamic Environment*, Kluwer Academic Publishers, 1998.

Baets W., The Hybrid Business School: a four-part fugue for business education, management development, knowledge management and information technology, Inaugural address, June 1999.

Baets W., *Une interprétation quantique des processus organisationnels d'innovation*, HDR, IAE Aix-en-Provence, 2004.

Baets W., *Knowledge Management and Management Learning*, Springer, 2005.

Baets W., Van der Linden G., *The Hybrid Business School: Developing Knowledge Management through Management Learning*, Prentice-Hall, 2000.

Baets W., Van der Linden G., *Virtual Corporate Universities: A Matrix of Knowledge and Learning for the New Digital Dawn*, Kluwer Academic, 2003.

Baets W., Venugopal V., "An IT architecture to support organizational transformation", in Galliers R and Baets W (eds), *Information Technology and Organizational Transformation*, Wiley, 1998.

Cohen J., Stewart I., *The Collapse of Chaos: Discovering Simplicity in a Complex World*, Penguin, 1994.

Langton C. (ed), *Artificial life, Santa Fe Institute in the Sciences of Complexity*, Proceedings Vol 6, Addison-Wesley, 1989.

Maturana H., and Varela F., *The Tree of Knowledge, The Biological Roots of Human Understanding*, Scherz Verlag, 1984.

Nicolis G., Prigogine I., *Exploring Complexity*, Freeman, 1989.

Prigogine I., Stengers I., *Entre le Temps et l'Eternité*, Fayard, 1988.

Stewart I., *Does God Play Dice?: The Mathematics of Chaos*, Blackwell, 1989.

Swan J., Scarbrough H., Preston J., "Knowledge Management – the next fad to forget people?", *Proceedings of the 7th European Conference on Information Systems*, Copenhagen, 1999.

Venugopal V., Baets W., *An Integrated Intelligent Support System for Organizational Learning – A Conceptual Framework*, The Learning Organization, Fall 1995.

PART 2

Academic Studies

Chapter 4

Value Creation through Intangibles: Emerging Good Practice

4.1. Introduction

The way value is created is changing. In many parts of the world better access to education and information is inexorably altering the behavior of societies and markets. Paradoxically, more knowledge and information also leads to increased uncertainty.

Organizations that wish to be effective – whether in public service or in commercial enterprise – are realizing that industrial era tools and techniques are no longer adequate. To survive and prosper they need to mobilize every resource at their disposal.

These trends are exposing the fact that many organizations possess a wealth of resources that have been taken for granted or allowed to lie fallow or even deliberately concealed. Such resources would include reputation, specialized expertise and intellectual property. At a deeper level lie organizational culture, values and beliefs.

Collectively, these types of assets are known as "intangibles" – but for many people the term is opaque and does not set the pulses racing. This chapter seeks to throw light on the subject; to reveal the importance and value of intangible assets and to show what practical steps can be taken to unlock their potential.

Chapter written by Nigel COURTNEY, Clive HOLTHAM and Chris HENDRY.

To a great extent the value of intangible assets depends on context. A deep understanding of biotechnology may be of little use to a hotelier. There is no "best way" to unlock hidden wealth. Each organization needs to identify and foster those hidden assets that it believes will sustain it in the future. Then they must be managed.

This requires ingenuity, commitment and, most importantly, communication. Recent business collapses are a stark reminder that the ultimate fate of an organization is usually determined by the perceptions of people outside it; regulators, experts, accountants, interest groups and the ordinary person in the street. This complex network of players is behind the rising pressure for greater corporate transparency, although, to be fair, not all these groups are equally enthusiastic about improved reporting of intangibles.

Pioneering organizations, institutions and governments have been exploring methods for dealing with intangibles for at least a decade. But there are still no mandatory requirements to measure and report such assets. UK plc does need to be aware that other EU states such as Denmark are gaining a significant lead. Forewarned is forearmed.

In the short term, progress is dependent on innovative practitioners. Those that are taking the initiative are doing so voluntarily. They are doing so because they know that the practice opens up great potential for business improvement and stimulates innovation. The management of innovation is usually promoted by advocates as good in itself. In reality, all innovation carries a degree of risk. This chapter does not portray the reporting of intangibles as risk-free. Rather, it argues that well-managed organizations can take steps to minimize these risks whilst harvesting the corresponding business benefits.

As evidence, we draw on and summarize the experiences of eight organizations that are serious about managing their intangible assets. They are variously based in the UK, other EU states and the USA, pursue different lines of business and vary in size from small to multinational. The common factor is that all are achieving outstanding performance. We conclude with a distillation of key messages that offers a menu of options to organizations that intend to emulate these thought leaders.

This chapter is derived from a Cass Business School research study sponsored by Messrs B&Q, entitled "Unlocking the Hidden Wealth of Organizations". This study [COU 03] by the authors in collaboration with Hill & Knowlton was presented at the UK Houses of Parliament in December 2003.

4.2. A remedy for our times

The new millennium has ushered in uncertainties that increasingly trouble the leaders and citizens of our world. Economies have experienced a dramatic boom and bust cycle, populations are ageing, pension systems are under stress, corporate scandals have shaken public confidence, the pervasiveness of computer-mediated communications has brought both benefits and burdens. Globalization of business, environmental pollution and random acts of terrorism have added to the sense of foreboding.

History tends to repeat itself. A century ago Rudyard Kipling's advice, if times were tough and if effective remedies were in short supply, was: "keep your head" [KIP 10].

Today's pressures and uncertainties have been forcing nations, markets and organizations to look to new sources for value creation. In recent years this has led to a quiet but steady transition from an exclusive focus on tangible assets towards a more holistic view of organizational resources. For example:

– OECD countries are annually investing 50–100% as much in the acquisition of knowledge as on physical assets [HIL 02]. US companies are annually investing approximately $1 trillion into intangibles [NAK 01];

– knowledge workers were the fastest growing segment of the OECD's labor force in the 1990s, growing at an annual average of 3% [OEC 00];

– by 1998 only 15% of the S&P500's market value was attributed to tangible assets, compared to 62% in 1982 [DAU 01]. Between 50 and 90% of the value generated by firms was attributable to intangibles [HOP 98].

For most of the 20th century, management good practice concentrated on maximizing returns from capital investment and physical resources. The economic and accounting systems supporting this approach shone a bright light on these resources through a tightly focused lens. But this harsh spotlight left other less tangible types of asset obscured in shadow. In the quest for improved performance, some organizations have been diffusing light across the wider landscape of their businesses and, in the process, illuminating hidden wealth that is locked up within their intangible assets.

In this chapter we set out to explain the nature and value of this hidden wealth and to describe practical steps for unlocking it. We draw on the experiences of eight organizations that have recognized that their intangible assets are a major source of wealth and are committed to identifying, developing and communicating the way they manage these assets. These beliefs are borne out by their willingness to collaborate in the research for this report and to allow others to share the lessons learned.

These thought leaders have decided to manage rather than be deterred by the risks that can accompany increased transparency. By "keeping their heads" they are deriving substantial business benefits.

4.3. Dispensing with the mystery

In ordinary conversation the subject of "intangible assets" rarely sets the pulses racing. For most people the nature of the topic is masked behind arcane management-speak. So let us look at what the management of intangibles actually means in more natural language.

The familiar meaning of "intangible" – as defined in the *Oxford Illustrated English Dictionary* – is: "that cannot be touched; impalpable". In an organizational context it is generally used to refer to assets exhibiting these qualities.

By convention, the term *tangible asset* is used to describe such things as property, buildings, plant and equipment, stocks and raw materials and financial capital. Traditionally, such assets, together with labor, have been treated as the fundamental economic resources and factors of production. Characteristically, they are things that can be measured or counted in accordance with generally accepted accounting principles.

In contrast, *intangible assets* are actual or potential resources that do not fit comfortably within these categories and, typically, are hard to measure in absolute terms. They exist across a spectrum ranging from intangible goods, through intangible competencies to latent capabilities:

– *Intangible goods and products*: these include licenses, franchises, patents and trade secrets. Broadly speaking, the value of such assets can be established in the marketplace.

– *Intangible competencies*: these include distinctive competencies, routines and business processes. Such assets are tradable in highly specific contexts but, typically, their value is most evident when it has been lost rather than when it has been acquired. In general, this type of asset exists within human beings and so the intrinsic value is established in the behavior of the possessors.

– *Latent capabilities*: these include such things as reputation, leadership and workforce caliber. The value of this type of asset is established in the perceptions and expectations of stakeholders. Therefore, this value tends to be manifested in indirect indicators such as the cost of capital and relative share price performance.

These examples of intangibles represent the tip of an iceberg of potentially valuable assets. The range of this hidden wealth reflects every organization's unique combination of history, purpose and beliefs. Despite this variety, long term research at Cass Business School, City of London, has found that organizations choosing to unlock their hidden assets tend to select from a relatively small number of "entry points" based on organizational priorities. This finding flows from Cass's involvement in:

– 1999–2001; membership of the European Commission's High Level Experts Group on the policy issues for the intangible economy. Final Report ISBN 92-894-0019-6;

– 2001–2003; lead partner in the EU-funded pan-European PRISM research project on the measurement and reporting of intangibles. http://www.EUintangibles.net;

– 2003; authorship of the BSI Guide to Measurements in Knowledge Management. PD7502.

The preferred entry points include intellectual property rights (IPR), brand value, knowledge, innovation and the reporting of intellectual capital.

The IPR route is a priority for relatively few organizations. This report seeks to elaborate the other four routes that may have broader utility. These routes need not be mutually exclusive but they are distinguished from each other as follows:

– *Brand value*: brand value is positioned as an intangible good. A focus on brand value provides an important entry point into the management of intangibles for organizations whose business activities also require substantial investment in tangible assets such as sales outlets and distribution capabilities.

– *Knowledge*: the term "knowledge" can usefully characterize the second category of intangible assets, namely "intangible competencies". A focus on the knowledge theme is particularly important to organizations whose product or service effectiveness is almost wholly dependent on the creation and deployment of knowledge to the right people at the right time.

– *Innovation*: innovation is a driver of future wealth. This theme is of particular importance to organizations that operate in hypercompetitive and rapidly changing markets. The continued existence of such businesses is determined by their ability to create or identify and respond to market opportunities. The capacity to innovate characterizes the third category of intangibles: "latent capabilities".

– *Reporting intellectual capital*: this theme tends to be adopted by knowledge-intensive businesses. Its importance is that it is a business process for communicating the effective management of the whole range of assets and explicitly shows how this practice enhances the business performance of the organization. As the resulting transparency exposes the organization to the public gaze, this theme is usually accompanied by thorough risk management procedures.

The EU-funded PRISM project, a two-year pan-European research effort that reported in September 2003 (see http://www.EUintangibles.net/research_results), has formulated a model of the new corporate asset base. This presents assets across a spectrum that highlights degrees of intangibility (see Figure 4.1).

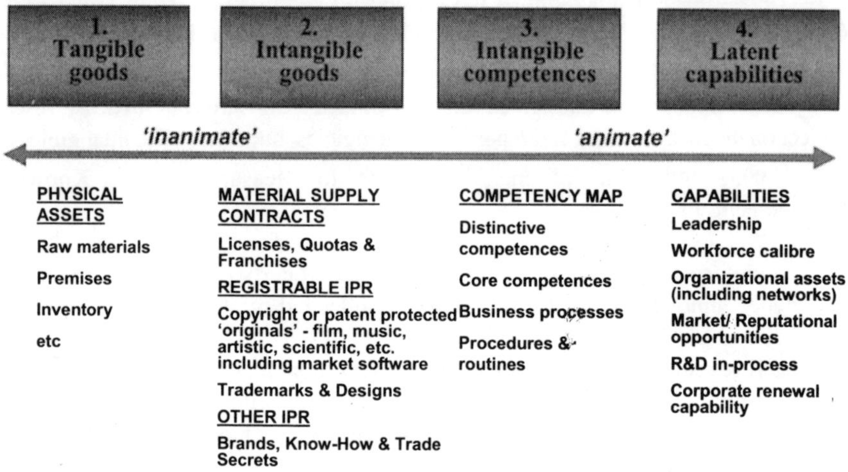

Figure 4.1. *The New Corporate Asset Base formulated by the PRISM project in 2001*

Over the two years of the project, PRISM researchers based in seven European states produced more than 100 reports and a themed collection of 15 in-depth case studies. Taken in the round, this body of work highlights the steady transition from an industrial era to a knowledge-based era – a transition that is already evident in developed economies and becoming apparent in rising economies such as those of India and China.

The research has identified, as evidence for this shift, the emerging organizational strategies and patterns of behavior that characterize good practice. The most salient of these include a tendency for organizations to foster formal and informal linkages with suppliers, customers and society at large – in other words, to become "networked organizations". This transition is accompanied by increasing awareness of the importance and value of corporate social responsibility and the management of knowledge.

The findings reflect a growing understanding that superior performance in the knowledge era requires managers to treat intangible assets as factors of production that are just as important as the industrial era factors. According to PRISM estimates, approximately 75% of the potential for value creation lies buried in intangible products, intangible competencies and latent capabilities. Organizational effectiveness in the 21st century will depend on the ability to create value through the managed interplay between tangible and intangible factors of production.

The idea is not new. Alfred Marshall (born in Bermondsey, London in 1842; died 1924) was a leading economist of his generation and, *inter alia*, established principles for the theory of value. Over 100 years ago he wrote: "Capital consists, in a great part, of knowledge and organization. Knowledge is our most powerful engine of production". Clearly, practical action to apply this wisdom is long overdue.

In its final report the PRISM project provides a diagram to illustrate how the managed interplay of tangible and intangible factors of production catalyses a value creation cycle (see Figure 4.2).

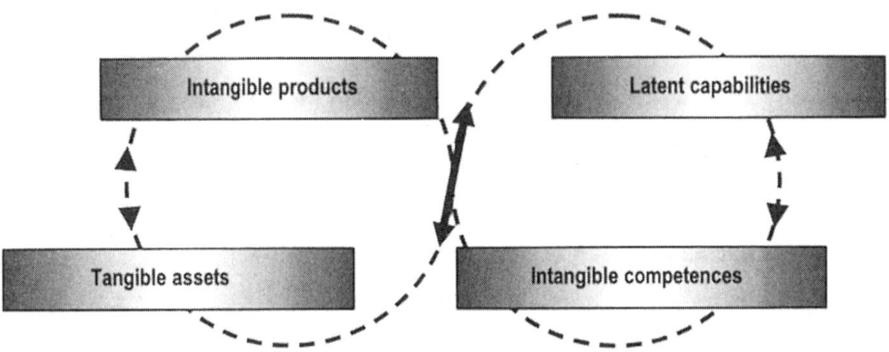

Figure 4.2. *The tangible/intangible value creation cycle – from the PRISM project 2003*

In practical terms, pioneering organizations in a number of industry sectors have been managing this interplay for some time. For example, in the pharmaceutical industry the value creation cycle tends to start with funding for Research & Development. When this leads to discovery of a novel compound it is followed by a patent application. If clinical trials of the new drug are fruitful, the quest moves to development of an economical production process. A successful drug reinforces the brand value of the organization and produces the revenues that fund new Research & Development … and the cycle begins again.

4.4. Value creation in the 21st century.

As any manager can testify, there is no "one right way" to ensure superior business performance. By the same token, few would deny that the socio-economic climate is being changed inexorably by the need to manage and balance process, technology and people-related issues.

These factors are reflected in the range of tools and techniques now available to managers. Since 1990 this toolbox has expanded to include, amongst others, Business Process Re-engineering [DAV 90], Information Systems Strategy and Applied Knowledge Management.

These successive waves of management innovation shared one thing in common. All required managers to learn how to harness intangible resources: the intellectual property of the organization, the value of its brand and reputation, the capabilities of new information and communications technologies, the knowledge and skills of people both within and outside the organization – including "the customer". And the managerial skill that links these factors systematically is communication.

Clearly this is a rich recipe. The possible combinations are many and varied; one size does not fit all. The goals and the circumstances of an organization will determine which of these intangible resources should be the chief focus of attention.

4.4.1. *Selecting an approach*

Many governments and supra-national bodies are investigating possible methods for reporting intangibles. At present none imposes any mandatory requirement. Organizations that are taking the initiative and reporting on their intangible assets do so voluntarily and because they are convinced that the practice delivers benefits.

Now that some 80% of activity in advanced economies is service-centric, thought leaders have increasingly drawn attention to the importance of non-financial, intangible factors of production. Since the mid-1980s various approaches have emerged for increasing attention to these. Three stand out:

1. the "balanced scorecard" approach which seeks to relate financial metrics and hard-to-measure factors concerned with human behavior and its outputs;
2. concern with the quality of life and the impact of human activity on the environment;
3. technology-based standards, such as XBRL, which apply a new "grammar" to data in order to facilitate collation and comparison and to highlight trends.

Each approach has sought to develop models that characterize sustainable good practice and voluntary codes that demonstrate adherence. Such efforts have been diverse and disjointed but there are signs of convergence around the concept of complementing corporate financial statements with social and environmental reports – Corporate Social Responsibility reporting (CSR). The purpose of this so-called "triple bottom-line" reporting is to give stakeholders a fuller picture of an organization's business performance.

In Europe, the EC issued CSR guidelines in 1992 and a voluntary code in 1995. The UN refined these as ISO 14000 in 1996. In 2001, the EU made social and environmental reporting mandatory for businesses operating in the financial services sector.

Vested interests have delayed widespread adoption. In Italy, for example, only 52 organizations publish environmental reports and a mere 30 issue social reports [ZAM 02]. The story is similar in the UK, according to a UK government survey report [DTI 03] entitled "SME-nvironment" of 8,000 firms representing the 3.7 million UK businesses employing fewer than 250 people. This found that although 73% felt there was a link between environmental good practice, increased sales and profitability and better customer relations, a mere 3% currently have an accredited environmental management system. However, nine out of ten firms said they would act to avoid prosecution if legislation was imposed.

Despite this tardiness to take up the acknowledged business benefits available to early adopters, some well-publicized incidents indicate that ordinary consumers are forcing both nation states and companies to take greater care of that most intangible and valuable of assets – reputation:

– In 2002 the Enron scandal destroyed the reputation of Arthur Andersen almost overnight and caused panic among the general share-owning public. This compelled the US government to rush through the Sarbanes–Oxley Act making directors personally liable for misleading business information (for details see http://www.AICPA.org).

– When Shell announced it would sink its obsolete floating oil rig, the Brent Spar, in the deep ocean, a widespread boycott of its products forced the company to dismantle the rig safely and at huge expense.

– The NIKE brand suffered a severe setback when it was revealed that its suppliers used child labor on low wages to make its expensive shoes and sports equipment.

4.4.2. *Taking a balanced view*

In the absence of official guidelines for reporting intellectual capital, the methods developed to date have been piecemeal. However, it is evident that, typically, these methods recognize the principles of the Balanced Scorecard [KAP 96].

Robert Kaplan and David Norton, who conceived this innovative management tool in the late 1980s, had observed that in the increasingly turbulent business environment a single-minded focus on financial performance did not guarantee long term prosperity, or even survival. They proposed that a business should monitor and balance performance in four business "perspectives" – financial, internal business processes, customers, learning and growth. The performance is gauged by means of metrics that communicate the targets to be achieved.

Such metrics or Key Performance Indicators (KPIs) for the financial perspective are well established and widely used – for example, "return on capital employed" (ROCE). The same does not apply to the other three perspectives. These are concerned with less tangible and non-financial aspects and each business will need to select KPIs that are relevant to its particular objectives.

Non-financial perspectives can be hard to measure in absolute terms. Early adopters of the Scorecard approach have found it useful to refer to Professor Karl-Erik Sveiby's Intangible Asset Monitor for inspiration (see http://sveibytoolkit.com). Some experienced practitioners include indicators that gauge performance simply in terms of an increase or decrease. Their advice is only to measure something if you intend to act on the results; then agree targets that are meaningful to those involved. For instance, a pharmaceutical company might decide that for its learning and growth perspective it was vital to monitor the proportion of its research staff holding a PhD. In many other contexts this metric would be irrelevant.

The issue of context is crucial. Different types of organization may employ a particular non-financial metric for completely different reasons. Take staff turnover rate for example: many organizations fear the loss of specialized knowledge and would want to keep this metric as low as possible. But firms in creative industries and consultancy are likely to prefer the rate to remain above 25% to ensure a constant influx of fresh talent.

4.5. Evidence of good practice

This section addresses practical applications. Over the past few years a small but increasing number of organizations have decided to give priority to unlocking their hidden wealth. In the absence of any nationally or internationally agreed standards or guidelines, these pioneers have been obliged to develop tailored approaches. Eight of these pioneers have agreed to share their experiences by way of the case studies that are appended to this section.

The cases have been selected to illustrate our four themes – brand value, knowledge, innovation, and the reporting of intellectual capital – with two cases for each theme. The common factor is that all eight organizations have outperformed during the difficult economic conditions since the turn of the millennium.

Of the eight subject organizations, six are private sector businesses, one is in the public sector and one is a public/private partnership. One operates solely in the UK, three are based in the UK, and of the remainder, one is based in Austria, one in Italy, one in Sweden and one in the USA. All except one employ more than 500 people.

The portfolio of detailed case studies accompanies the full report. A synopsis of the salient issues and lessons from the cases follows for each theme.

4.5.1. *Theme 1: brand value*

The importance of brand value is that it underpins the marketing strategy of the organization and communicates its values and beliefs to all stakeholders. A brand transcends the life-cycle of specific products and services and allows these to be extended, repositioned or replaced seamlessly. This enables sustainable growth and future revenues that, in turn, reinforce brand value. Business takeovers show that when brands are sold or licensed, they typically bring far greater proceeds than the tangible assets.

The featured organizations are B&Q and Whitbread.

4.5.1.1. *B&Q*

Since its humble beginnings as a single store in Southampton in 1969, B&Q has expanded both organically and by acquisition, and upheld its brand essence throughout. Brand value is one of the most intangible of assets because it is largely determined by people outside the organization. B&Q has built a multi-level approach to brand development that is designed to create and return new value from investment.

Alongside its brand development strategy, B&Q has pioneered a range of complementary initiatives for nurturing its other intangible assets, including e-learning via the B&Q University and B&Q Social Responsibility. Innovation of both brand and product deliver additional intangible value to B&Q and its parent company Kingfisher in the form of reputational capital.

The principal lessons from B&Q's experience are as follows:

– Fostering brand value engenders a widespread public perception that the company stands for quality, style and innovation.

– A focus on CSR leads to greater business efficiency and raises the reputation of the company among staff and in the minds of existing and potential customers.

– The policy of measuring and reporting on the management of intangibles has an impact on the supply chain. B&Q's suppliers have responded with, for example, radical improvements in forestry, a substantial reduction of volatile chemicals in paints and the development of peat-free growing media.

– Intangible values can be difficult to measure. "Proxy measures" such as regular independent surveys of staff and customer satisfaction can provide powerful indicators of any increase or decrease in value. These indicators alert managers to take early action that will reinforce financial results.

4.5.1.2. *Whitbread*

After 250 years as a leading brewer, Whitbread became one of the first UK companies to sense and respond effectively to the formation of what is now the leisure industry. During the 1990s, the company embarked on the bold strategy of divesting its beer production and distribution facilities and re-investing in a portfolio of high profile brands, including hotels, restaurants and health & fitness clubs. The company has embedded its original values and beliefs into a "balanced scorecard" approach and the successful transformation has been underpinned by the use of brand "templates" that set out the structure and discipline for managing each brand effectively.

The company's published results show how the brand portfolio strategy has resulted in a broad-based platform providing stability, resilience and growth. Between October 2000 and August 2003, only half of FTSE 100 companies delivered any growth in total shareholder returns; thus Whitbread's achievement of 80% growth in this period places it in the elite Top 10.

The chief lessons learned by Whitbread are as follows:

– Effective brand management follows a six-stage cycle that starts and ends with "consumer insight". This enables trends to be anticipated and brands to be differentiated.

– In the leisure industry, brand value is indissolubly linked to both customer and employee satisfaction. Key to this is a method for managing such intangible factors of production that is understood by all and is clearly linked to the reward system.

– Regular customers increase brand value, so a consistently satisfactory experience is vital. The management of intangibles must therefore extend beyond corporate boundaries to suppliers and other alliance partners.

– A clear link between intangible assets and "the bottom line" is crucial when growth opportunities beckon. Whitbread's ability to manage intangible assets has delivered over 80% growth in total shareholder returns since October 2000 – placing it in an elite Top 10 among FTSE100 companies and fuelling ambitious plans.

The processes these two companies have developed for tracking and fostering brand value are notable for addressing customer and employee satisfaction, and for linking both to business performance outcomes.

4.5.2. *Theme 2: knowledge*

The importance of knowledge is that it is the essential ingredient for creativity and drives organizational effectiveness. 400 years ago, Francis Bacon proclaimed that "knowledge is power". Today, organizations are increasingly recognizing that the value of knowledge increases exponentially when it is shared and accessible at the moment it is needed.

The two featured organizations are Bloomberg of New York and the UK Fire & Rescue Service.

4.5.2.1. *Bloomberg*

In 1981, Mike Bloomberg launched an offering that, until then, had been impossible – real time bond market data supplied digitally and combined with analytical tools for application by the client. The service gave early adopters such competitive advantage that the firm achieved a global presence within six years. Continuous innovation and extensions to functionality have driven sales growth to become exponential. Today, Bloomberg is the leading provider of financial market information and one of the world's most received radio, TV and web broadcasters and publishers.

With over 8,000 staff in 94 offices worldwide, the concept of workspace is fluid. There is no organogram and there are no job titles. People learn and develop the business and themselves by frequent relocation to cross-fertilize ideas and boost management skills. The company's own product is the backbone of the organization. Instant messaging has virtually eliminated paper. The system tracks and records all activity. This provides staff with ready access to structured information that, in any given context, unleashes the corporate knowledge of the organization.

The chief lessons Bloomberg has learned are:

– very rapid growth is only sustainable if the company becomes a “learning organization”. When employees are aware of and can contribute to the knowledge and competencies that exist across the enterprise, they can grow with the business;

– a tightly defined organizational structure can inhibit knowledge-sharing and the development of experience. It is more important to be able to reach people with know-how than to know what they do in their job role;

– openness and transparency demand loyalty and mutual trust, Bloomberg is a private company. People who join are treated as family and developed. Those who leave can never return.

Experimentation should be encouraged and failures forgiven. It is the synergy between business units rather than the contribution of each to the “bottom line” that leverages sales of the core products.

4.5.2.2. *The UK Fire & Rescue Service*

The visible side of fire service work is dramatic, heroic and palpably tangible. This case throws some light on the hidden side – a side that is strongly focused on the effective management of intangible assets such as knowledge, skills and safety in order to achieve the economical delivery of superb public service.

As part of this, the National Mobile Data Project is transforming the way in which relevant information and collective experience are being captured and made accessible in real time to officers at the fire ground and to other emergency services. This work has already been independently assessed as of national importance.

The principal lessons learned by the UK Fire & Rescue Service are:

– when crucial business decisions must be made away from the office, the organization stands or falls by its ability to deploy relevant knowledge and information in real time;

– organizations possess vast amounts of knowledge, but much of it resides in people’s heads. Its essential documentation is an expensive process. The investment is worthless if knowledge cannot be readily and remotely accessed and used when required;

– when time is of the essence, the user needs access, in the first place, to the most immediately relevant information. Therefore, the organization should audit all its information and knowledge and then decide which is most important to the achievement of its objectives;

– information provision must be geared to the financial and physical resources that enable knowledge and information to be acted upon. Managers must prioritize the more efficient utilization of property and equipment.

Both Bloomberg and the UK Fire & Rescue Service recognize the importance of knowledge in the business, and for information to be organized so it can be accessed in a timely and relevant way. Well-structured information frees up people to act flexibly in accordance with the demands of the situation.

4.5.3. *Theme 3: innovation*

The importance of innovation is that it is a process that enables concepts and ideas to be converted into products and services that people will want. Existing products can be copied by rivals or can quickly fall out of fashion. In many industry settings, the capacity to innovate is a prerequisite for survival. This capacity is a function of attitudes and behaviors that are understood and shared across the enterprise.

The two featured organizations are Intercos of Milan and mmO2.

4.5.3.1. *Intercos*

Milan-based Intercos is a specialist in color cosmetics. Each tonne of the talcum it mines makes over 100,000 items of make-up – in any color imaginable – for the world's best-known beauty brands. The transformation takes 16 weeks and the added value is enormous … but fashions change overnight. Innovation is the life-blood of the company and the way through which Chairman Dario Ferrari envisions Intercos becoming the world's leader in the color cosmetics field: "We have no other strategy than continuous innovation".

Intercos has embraced the management of intangible assets, in itself an innovative approach, as the principal driver of business growth through client relationships. The development of an Intangible Capital Statement to communicate the company's values has been widely perceived to be of strong importance, with a remarkable impact on customers who view the decision to show Intercos in full transparency as enlightened.

The chief lessons learned by Intercos are as follows:

– there is no such thing as a generalizable tool for managing intangible assets because every organization has a unique set of value creation drivers. At Intercos, the main source of value creation is innovation; the driving force resides in its customers;

– the application of an Intellectual Capital (IC) approach means radical change to the company's culture. The commitment of top management is essential to secure the buy-in of all stakeholders. If an IC report is produced merely as a PR exercise, it will almost certainly backfire;

– an IC report exposes weaknesses to the public gaze; the report must state clearly what action is being taken. The pay-back comes when the value creation focus moves from measurement to management;

– in Intercos's experience, its Intangible Assets Report supports and justifies its policy of re-investing 24% of net sales revenues in new product development. Intercos has increased this every year since 1994 and quadrupled sales during the same period.

4.5.3.2. *mmO2*

The fostering of intangible assets is playing a vital part in the establishment of mmO2 – the listed, re-launched company derived from BT Wireless in 2001 – as a credible, innovative, independent business. mmO2 concludes that the assiduous management of intangible assets has helped it post a sound financial performance for seven successive quarters and convince many in the City that it is a serious company.

The group's first annual Corporate Responsibility report was published on the mm02 website in July 2003. In October 2003, results showed that company is on track to reach 30% operating margin for the year and achieve doubt-digit growth.

The chief lessons that mmO2 has learned in implementing this policy are as follows:

– CSR is a company-wide way of thinking, which focuses attention on problems to be profitably solved and new opportunities to be pursued;

– CSR identifies key stakeholders and issues the company should engage with, and brings it closer to its customers;

– as a result, CSR becomes a management tool for motivating staff, dealing promptly with problems, lapses and omissions, and communicating with customers and investors. In short, it leads to innovation;

– tracking company performance on CSR measures makes everyone aware of the importance of "intangibles" as the source of business success in a climate where traditional sources of earnings can no longer be relied upon.

These two companies use different reporting tools to reflect intangibles, but both have the effect of focusing employees on customers as the driving force for innovation.

4.5.4. *Theme 4: reporting intellectual capital*

Reporting intellectual capital is a business process that communicates the effective management of the whole range of assets. It presents goals, targets and results regarding the organization's people, its organization and its customers, and explicitly shows how this relates to and enhances financial performance. The resulting transparency carries risks that are outweighed by stakeholder confidence. IC reporting is of particular utility to knowledge-intensive businesses whose success will derive from the potential value of intellectual property, know-how and workforce caliber.

The featured organizations are Austrian Research Centers of Seibersdorf and Celemi of Malmö.

4.5.4.1. *Austrian Research Centers (ARC)*

As Austria's largest Research Technology Organization, ARC's main function is to manage the transfer of knowledge between state-owned laboratories and universities and the applied research and development in companies. The chief outputs of this public–private partnership are highly intangible: knowledge that can be re-applied and workable innovations that can be made into products. In short, ARC's business is to foster intellectual capital (IC).

ARC decided that traditional accounting procedures did not allow it to report adequately to stakeholders on the management of its IC. In 1999, ARC developed a new approach to measuring and reporting IC based on transparency about goals and business strategies, performance indicators and data-gathering techniques, change management issues and project management procedures. ARC has since established this innovative, process-oriented method as an internal management tool and for communicating with other stakeholders in a supplement to the annual financial accounts.

The principal lessons that ARC has learned are as follows:

– the company's mission should be expressed in terms of its primary "knowledge goals". The business processes that underpin their achievement can then be identified;

– an organization must decide which types of intellectual capital (IC) are important to it and should be measured for each knowledge goal. ARC measures, balances and reports five types – human, structural and relational capital, key processes and results;

– as an intangible asset, knowledge can be replicated many times. Any IC reporting model must ensure that an intangible asset is not double counted. This requires people across the enterprise to understand and actively participate in the measurement process;

– an effective IC model provides both an internal management tool and a means of communicating with external stakeholders. It should present precise definitions, numerical values, qualitative explanations and symbols depicting results trends and intended actions.

4.5.4.2. *Celemi*

The measurement and reporting of intellectual capital is still evolving as an essential business activity, despite clear bottom line business benefits. In the early 1990s, Celemi, a Swedish learning design consultancy, developed a business simulation that enabled executives of knowledge businesses to understand that financial and intellectual assets must be managed equally well.

Believing it should "practice what it preached", Celemi applied the methodology to itself. The resulting Celemi Monitor reveals fully the non-financial metrics that drive performance. The Monitor has been published alongside the company's financial statements since 1995. Notably, the Monitor helped Celemi overcome problems during the e-commerce downturn, restoring stability and dramatically improving profits. This small company now operates in over 20 countries, serving many of world's leading organizations.

In Celemi's experience, the chief lessons are as follows:

– the monitoring of intellectual assets highlights the non-financial metrics that drive business performance. It signals trends requiring attention and enables staff to participate in decision-making and resource allocation;

– start with measures that lead to increased customer satisfaction and follow with measures that nurture employee satisfaction. This provides the basis for reducing costs. The secret is simple – do not measure activity; measure the results of activity;

– financial statements reveal past performance. The measurement and reporting of intangible assets exposes how the business is positioned for the future and focuses attention on the things that will deliver tomorrow's profits;

– publication of an intellectual assets report is unlikely to affect clients' purchasing behavior. Rather, it enables the company to identify and develop the right sort of clients – those that enhance the company's image and buy regularly because they are attracted by the company's thinking, not simply its products.

The development of intellectual capital reporting is a demanding task, but it pays off. The two cases highlight its value in revealing the underlying drivers and processes for future business performance. It helps these organizations to think deeply, and to think ahead.

4.6. Key messages

In advanced economies manufacturing now represents less than 20% of gross domestic product. The balance has swung in favor of service industries that supply intangible goods, competencies and capabilities. And consumers of manufactured goods now expect and demand knowledge to be embedded in the products. The pervasiveness of information and communications technologies has transferred power to consumers and helped replace "supplier push" with "customer pull".

However, economic, accounting and management systems have not kept up with the pace of change. Good practices perfected during the industrial era are inadequate for reporting performance in an intangible economy.

Against this background, thought leaders in a variety of industry settings have been quietly developing methods and processes for identifying and enhancing their intangible assets. Instead of just shining a spotlight on financial performance, these pioneers have been diffusing light across the wider landscape of their businesses and exposing valuable assets that have been neglected and allowed to lie dormant. This illumination also reveals connections and pathways that can be cleared to facilitate the interplay between intangible and tangible assets.

This report has highlighted the experiences that eight of these pioneering organizations have been happy to share. The following key messages stand out:

– *Context*: whereas financial accounting is based on generally accepted rules that allow stakeholders to readily make inter-company comparisons, there are presently no agreed standards for measuring and reporting intangibles. Accordingly, interpretation of IC reports requires an understanding of the contexts in which the reporting organization operates.

– *Value creation*: explicitly managing brand value focuses the attention of stakeholders on the ways in which the organization creates value. This not only increases internal effectiveness but also enables supply chain partners to contribute to improvements for mutual benefit.

– *Risk*: the reporting of CSR or IC exposes both the strengths and weaknesses of the organization to the public gaze. The reporting process should therefore be accompanied by thorough risk management procedures, and reports should clearly state planned actions.

– *Innovation*: CSR and IC reporting is a company-wide way of thinking that promotes innovation and creativity. Systems must be in place to manage, reward and capitalize on such behaviors. Initiatives designed to present a favorable public face are likely to backfire.

– *Communication*. The management of intangibles creates a potent channel for communicating with stakeholders and for motivating staff. It is not just a medicine;

it is more like a corporate fitness regime that prepares the organization to withstand new pressures and threats.

– *Objectives*: to secure the active participation of staff the organization's mission should be expressed in terms of explicit knowledge goals. Key performance indicators can then be selected to show whether or not each goal is being achieved.

– *Performance*: the effective management of intangibles builds on past performance by causing employees to think about improvements and innovations that will address and deliver sound performance in the future.

These key messages constitute a menu of policy options that can be used individually or in any combination. They are offered to businesses that wish to emulate thought leaders in the management of intangibles and unlock the hidden wealth in their organizations.

4.7. About the authors

Nigel Courtney is an honorary visiting fellow at Cass Business School, City of London, where he gained his PhD and lectures on e-business and Applied Knowledge Management. He is also a visiting research fellow at the Australian Graduate School of Management, Sydney. Nigel is a chartered engineer with extensive experience in project and general management. Through his firm, Courtney Consulting, established in 1990, he advises clients on information systems strategy and the management of intangible assets.

Clive Holtham is Professor of Information Management at Cass Business School. He trained as an accountant and, following six years as a Director of Finance and IT, he moved to the Business School in 1988, where he researches the strategic exploitation of information systems and knowledge management. He managed the EU's €1.6m study PRISM (Measurement and Reporting of Intangibles).

Chris Hendry is Centenary Professor in Organizational Behavior, Associate Dean (Research), and Director of the Centre for New Technologies, Innovation and Entrepreneurship (CENTIVE) at Cass Business School. He is the author of some 15 research reports and monographs, 50 articles and book chapters, and three books. He has extensive research interests, including current projects on fuel cells and on the measurement and management of intangible assets.

4.8. References

[COU 03] Courtney, N., Holtham, C., Hendry, C. (2003) *Unlocking the Hidden Wealth of Organisations: The Development and Communication of Intangible Assets*. Printed by Hill & Knowlton for selected circulation. PDF version available for free download at http://cass.city.ac.uk/centive/news/index/html

[DAU 01] Daum, J. (2001) Business Management in the New, New Economy – How to Exploit Intangible Assets to Create Value. In proceedings of SAP's *European mySAP Financials Conference*, June, Basel, Switzerland

[DAV 90] Davenport, T., Short, J. (1990) "The new industrial engineering: information technology and business process redesign". *Sloan Management Review*. Summer: pp. 11–27

[DTI 03] UK Department of Trade & Industry (2003) *SME-environment Survey*. Published in September on http://www.netregs.gov.uk

[HIL 02] Hill, P., Youngman, R. (2002) The Measurement of Intangibles in Macroeconomic Statistics. PRISM research paper #5.4. Available for download at http://www.euintangibles.net/research_results/WP5/index_html

[HOP 98] Hope, J., Hope, A. (1998) *Competing in the Third Wave: The Ten Key Issues of the Information Age*, McGraw-Hill

[KAP 96] Kaplan, R., Norton, D. (1996) *The Balanced Scorecard: Translating Strategy into Action,* Boston MA: Harvard Business School Press. Explanatory information is readily available at http://www.bscol.com/bscol/leadership/

[KIP 10] Kipling, R. (1910) "*If–*" .Published in his anthology *Rewards and Fairies*.

[NAK 01] Nakamura, L. (2001) What is the US gross investment in intangibles? Federal Reserve Bank of Philadelphia; working paper # 01-15 and in proceedings of the *4th Conference on Intangibles*; New York University, May

[OEC 00] OECD (2000) Links between Policy and Growth – Cross-country Evidence. Draft paper for working party 1, Economics Department

[ZAM 02] Zambon, S., Cordazzo, M. (2002) *IC Statements vs Environmental and Social Reports: An Empirical Analysis of their Convergences in the Italian Context* (University of Ferrara). This paper can be downloaded from the PRISM online library at http://www.euintangibles.net/research_results/WP4/index_html; paper # 4.15

Chapter 5

Learning-by-Doing Knowledge Externalization: From Boundary Objects to the Emergence of Tacit Knowledge

Knowledge management (KM) has been defined as the systematic and explicit management of knowledge-related activities, practices, programs and policies within the organization and aimed at supporting the enterprise's ultimate objectives (Wiig, 2000). Following Hansen *et al.* (1999), we can map the KM field along two lines: managing explicit knowledge through Information Technologies, in the expert organization, and managing tacit knowledge through people, in the organization of experts. In their seminal work, Nonaka and Takeuchi (1995) develop a theory of knowledge, describing the interplay between tacit and explicit along four modes of knowledge conversion (see Figure 5.1): socialization (from tacit to tacit, from individual to individual), internalization (from explicit to tacit, from organization to individual), combination (from explicit to explicit, from group to organization) and externalization or explicitation (from tacit to explicit, from individual to group).

Chapter written by Jean-Michel VIOLA and Réal JACOB, HEC Montréal.

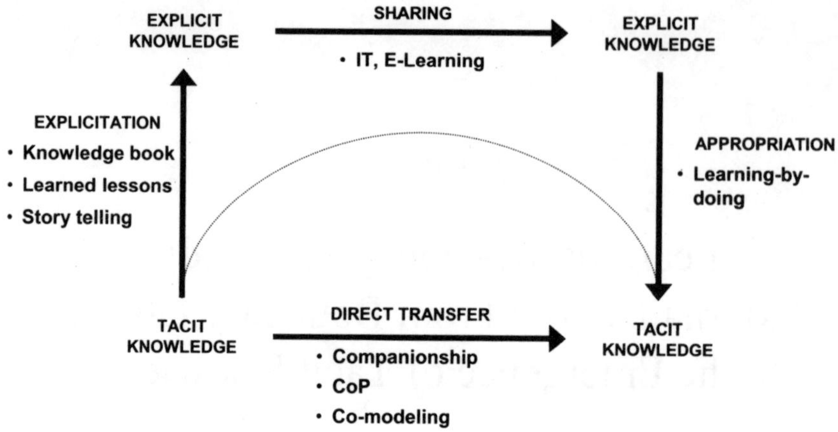

Adapted from Nonaka & Takeuchi (1995)

Figure 5.1. *Knowledge conversion in four modes*

Some major drawbacks have been identified in the process of externalization. More often, it has been seen as an artificial process, ignoring people and social dynamics. Amin and Cohendet (2004) discussed the idea that the externalization process "is a process of knowledge creation that alters both the codified and the tacit forms of knowledge" (p. 24). Following this statement, this text develops the idea that the process of explicitation itself co-creates both explicit and tacit knowledge and sheds light on two main issues: the critical importance of boundary objects and the need for a more precise definition of tacit knowledge.

After describing a case of externalization in a large Canadian industrial firm, we illustrate how this process of externalization has generated relevant and robust tacit knowledge as a side product. Thus, the rise of the expert organization may also foster tacitness. This scenario of co-occurrence raises striking questions to be addressed in many KM programs.

5.1. Learning-by-doing KM

A few months ago we were puzzled by a comment from a human resources director. Speaking about why her firm was engaged in significant KM research with academics, she said straightforwardly: "We could have hired consultants but they are learning with us at the same time… everybody is learning in this field! So we are more comfortable if it's crystal clear from the start!" This quote opens a large window of opportunity for action research at the same time as it illustrates a certain pragmatic approach to KM, a way to harness the value of experience in the knowledge-driven firm (Geisler, 1999). We can encapsulate this approach in the concept "Learning-by-doing", which means in fact three interelated features:

– *Learning over time*: being able to learn in action, being able to look back on the output of knowledge initiatives, to make room for adjustments. However, it is far from easy to do this (Huber, 1991) because you need to develop a sense of experience.

– *Learning together*: being able to connect individual experiences through collective thinking, as shown by the growing popularity of communities of practice, e-learning activities, story telling, learning organizations and corporate universities.

– *Learning from facts and from people* through corporate cases[1] and lessons learned[2].

We have chosen to present here the case of Power Corp (PC) to illustrate how the process of externalization, rooted in the will to hedge the risk of knowledge leakage, co-creates tacit knowledge. We emphasize how the building of knowledge artifacts, acting as boundary objects, foster tacit knowledge.

5.2. A process of externalization: knowledge strategy at Power Corp

PC is a large Canadian industrial firm providing services in the energy sector. Net income totaled over 1b euros in 2003. PC has formalized its knowledge strategy into a *strategic KM plan* (2004–2006). This is an ongoing project largely focused on human resources in the context of a specific threat: the expected retirement of thousands of baby boomers in the next few years. The plan is then specifically targeted toward hedging of the risk of a potential loss of expertise. Then externalization issues are critical in this context.

---

1 Siemens, with the help of Tom Davenport and Gilbert Probst, develop a company-wide approach to case building and transferring. In the pioneer book, *Knowledge Management Case Book* (Wiley, 2000), 13 cases about innovative KM practices, written by Siemens managers, are presented with key propositions and discussion questions.

2 Learned lessons methodologies was first developed on a large scale in the US army through the CALL (Center for Army Lessons Learned). For more details on the approach, see, for example, Baird *et al.* (1997).

Nonaka and Takeuchi (1995) define externalization as the process of articulating tacit knowledge into explicit knowledge. They stress the fact that this is triggered by dialogue and reflection and that it represents the most intensive interaction between tacit and explicit knowledge from individual to group entities. The new competitiveness of organizations is based on collective knowledge and, from that standpoint, externalization is the key process. In this way, externalization holds the key to knowledge creation, taking the sequential shapes of metaphors, analogies, concepts, hypotheses and models. PC develops such a sequential four-phase externalization process through its own specific activities: knowledge mapping, criticality assessment, knowledge modeling and knowledge planning. Outside the sphere of marketing, Western firms are not very good at playing with metaphors and analogies, despite the seminal work of Morgan (1991). They are much more at ease with concepts, hypotheses and models. Contradictions and fuzziness, usually associated with metaphors and analogies, are very often seen as errors and a lack of control, rationality or order.

At PC, the KM strategy is rooted in two complementary perspectives: first, to better understand what are and what will be our key needs in terms of knowledge, second, to favor the acquisition, development and transfer of such knowledge so as to avoid threats to future performance. Thus, KM at PC is organized around four lines (see Figure 5.2, below) to take risk hedging actions: knowledge mapping, criticality assessment, knowledge modeling and knowledge planning. In this chapter, we mainly describe the process of externalization as the basis of such a knowledge strategy.

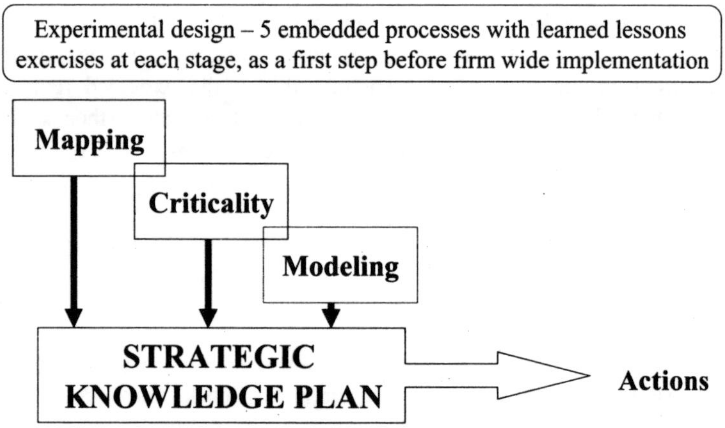

Figure 5.2. *Knowledge strategy at Power Corp*

Below, we describe each step in more detail.

5.2.1. *Mapping as a knowledge audit and mapping as scenario planning*

First, a knowledge mapping process started in the R&D division. The six directors organized brainstorming sessions around one question: what are our expertises? After a few iterative rounds they reached an agreement on a graphical map linking expertise from general knowledge to its applicability within the firm. The process was seen and actually carried out as a knowledge audit. But despite the fact that they gathered key information, it was merely information and in no way was it a basis for decision-making in terms of knowledge planning. They needed some sort of quantitative estimates of the real value of what they know, now and in the future, over a five to 10 year time frame.

5.2.2. *Criticality assessment scales as decision rules*

In this context, it was decided to apply at PC the methodology once developed by Jean-Louis Ermine and the Club de Gestion des Connaissances in France (Ermine, 2003; Boughzala and Ermine, 2004). The organization started a process, lasting nearly two months, through over 60 interviews with people to better evaluate the criticality of what they know. It used 21 criteria along four axes (utility of knowledge, rarity of knowledge, and ease of capitalization and ease of access). This process led to two types of results:

– a substantial description of the knowledge portfolio and trends;

– a qualitative description of the nature of the knowledge and the needs for transfer.

These two outputs were the basis for the next step:

– decisions on knowledge portfolio: hold, cut, develop alone or in partnership, etc.;

– decisions on the need for and the nature of knowledge transfer: what should be prioritized in terms of transfer? How should it be transfered?

5.2.3. *Modeling as a learning tool, as co-creation of knowledge*

Modeling is used at PC in two very different ways:

– the classical modeling approach to writing the book of knowledge, procedures or methodologies to guide future work;

– a more innovative approach using co-modeling in the case of a very complex task. Using simple modeling software, both the expert and the novice engage in a modeling process concerning a specific task: the expert models his task, while the novice models what he understands of the expert's task. Then they engage in a co-construction of the model until they are both satisfied with the representation. This leads to a valid description of the expert knowledge and it is also a description that is understandable from the novice's point of view. This way of doing things has proved to be highly effective for complex tasks at PC.

5.2.4. *Plans as road maps addressing more specifically the need for tacitness*

Finally, a knowledge plan was written at the directors' level and then integrated at the level of divisions. The plan used different inputs based around a similar template: maps, criticality, human resources and workforce demography, trends, transfer plans. Four additional pilot cases were chosen to develop this process. After each level, in each of the five pilot cases, "lessons learned" research was carried out to identify a return on experience that could be reused. What we will discuss now is the co-creation of tacit knowledge that occurred throughout this whole process.

5.3. The tacit output of externalization: the importance of boundary objects

What is tacit knowledge and how is it created and nurtured? Since the seminal work of Polanyi (1958) and, more specifically, since the intensive use of the concept in the growing field of KM (Nonaka, 1994; Nonaka and Takeuchi, 1995), tacit knowledge has been widely discussed in the literature. Despite this abundance and the fact that it is considered a key issue in knowledge creation and dissemination, it remains largely a black box. Part of this elusiveness is due to its true nature, but part of it is also due to a lack of a comprehensive description of tacit knowledge at the individual level. In Polanyi's view, knowledge has a personal dimension and is context-specific. Human beings are then creating knowledge by involving themselves with objects and they know more than they can tell. Following Polanyi, Nonaka and Takeuchi (1995) described the trajectory from tacit to explicit and back to tacit, through socialization, externalization, combination and internalization. In their view, the creation of tacit knowledge is first an individual process. At the same time, they recognized (p. 69) that some kind of externalization helps to develop tacit knowledge:

> For explicit knowledge to become tacit, it helps if the knowledge is verbalized or diagrammed into documents, manuals, or oral stories.

The case study of KM at PC sheds lights on two issues: the critical importance of boundary objects and the need for a more precise definition of tacit knowledge.

5.3.1. *Externalization creates critical boundary objects*

Boundary objects (BO) are artifacts of knowledge that serve as both containers (content) and carriers (process) of knowledge (Grey, 2002), "being both plastic enough to adapt to local needs and constraints of the several parties employing them, yet robust enough to maintain a common identity across sites" (Star, 1989).

Then, at PC, we can consider knowledge maps, criticality scales, models and plans as BO, the mere existence of which fosters the emergence of tacitness.

These BO allow for combination, that is, according to Nonaka and Takeuchi (1995), the process of systemizing concepts into a knowledge system, combining different bodies of explicit knowledge. This combination creates redundancy, a key prerequisite for collective learning, the basis for collaboration and synergy, although redundancy is also seen as inefficient in the machine organization. Redundancy creates bridges for joint projects at PC. Moreover, the process of building artifacts and the artifact itself create overlaps, "learning by intrusion" (Nonaka & Takeuchi, 1995) in others' sphere of expertise.

Then, maps, criticality scales, models and plans effectively make KM happen in two ways:

– by fostering socialization. Artifacts, as referentials, create a shared space that is the basis for dialogue (Schrage, 1995);

– by fostering internalization, creating identity and meaning. The fact that people are now able to represent themselves in the knowledge map brings some sense of belonging. Although this was not seen as an issue at the beginning, it led to some unexpected empowerment.

Hence, BOs were used by members in very different ways, although the representation is shared. As Grey (2002) stated, BOs are an important class of knowledge artifacts as they are associated with process, meaning, alignment and reification. They are center stage in the dynamics of knowledge exchange. Thus, they are key elements in fostering the emergence of tacit knowledge.

5.3.2. *Refining the classical definition of tacit knowledge*

To better understand the emergence of tacit knowledge through the process of externalization, we need to refine the definition of tacit knowledge. Starting from the classical distinction first made by Polanyi between tacit and explicit, Nonaka & Takeuchi (1995) tried to develop more explicitly the definition of tacit knowledge. They argued that tacit can include two dimensions:

– a cognitive element, a working model of the world;

– a technical elements, such as know-how, craft and skills.

Baumard (1999) brought the explicitation of tacit knowledge one step further by making another distinction between tacit knowledge that cannot be articulated or stabilized and tacit knowledge that can be. But the definitions of both Nonaka and Takeuchi (1995) and Baumard (1999) merely concern epistemological issues about tacitness. To consider tacit knowledge in-action, beyond know-how and skills, it may be useful to consider management activities implied by tacitness. Hence, we can look at BOs as fostering tacitness along four managerial dimensions: tinkering, judgment, connectivity, and coordination.

5.3.2.1. *Tinkering*

Tinkering refers to "do-it-yourself", the ability to invent, to put the parts together, to see how the parts fit so as to build innovation or new synergies. BOs allow for identification of parts in the organization, creating unexpected connections, and establishing ground for new ideas. At PC, BOs were seen as road maps for navigating through the whole organization and as tools to foster knowledge brokerage between knowledge components in order to develop innovative projects.

5.3.2.2. *Judgment*

Judgment is the ability to take decisions in action, shaped by experience. For example, judgment allows the salesman to be able to understand and to adjust to the demands and reactions of his customers. BOs bring a better evaluation of the decision-making context, improving judgment (Augier *et al.*, 2001). At PC, BOs were seen as collective referentials within groups and as building key bridges between groups. They allow for the refinement of mental models, improving the decision-making process by decreasing perceptual biases between groups within PC's complex organization.

5.3.2.3. *Connectivity*

Connectivity refers to the ability to map, understand and use the social network, to promote dialogue based on a shared space. BOs help to frame this network and to understand better the perspectives of other stakeholders. At PC, there is some concern about non-technical competences. Behavioral features are key and very often implementation issues raise behavioral problems rather than technical ones. And maps, criticality and models help develop transverse understanding, the ability to interact between divisions.

5.3.2.4. *Coordination*

Coordination is close to tinkering as it is also a combining capability, but here it is targeted toward operational efficiency and not toward creativity and innovation. It is rooted in Polanyi's example concerning the way a child is able to ride a bike, not using all the technical knowledge involved but simply by an appropriate coordination of his or her moves. BOs improve the efficiency of coordination by providing keys and identifying modular components. At PC, BOs are seen as a common language allowing collective and collaborative work.

The process of externalization produces boundary objects as output. In the case of PC, such BOs are knowledge maps, criticality assessment grids, knowledge models and KM plans. These BOs are knowledge artifacts that are fostering the creation of tacit knowledge as a side effect in the process of externalization. More specifically, they allow for faster innovation through improved tinkering capabilities, for better decision making through more accurate judgment, for better social interactions through improved connectivity, and for better operational efficiency through improved coordination.

5.4. Conclusions and lessons learned

Externalization and BOs increase both explicit and tacit knowledge. Although this process is seen as a valid framework from the actors' point of view, a lot of questions remain unanswered. And there are also some intriguing pathological sides to externalization.

Many hurdles still exist in the process of knowledge explicitation. First, it can be seen as being on the edge of bureaucracy through excessively formalized methodologies; second, the proof of time needs to be addressed as it is a long and costly exercise that needs to be carried out over time with a constant rigor; Third, it can be seen as a political issue, a way to build legitimacy around tough questions about downsizing or reorientation; Fourth, it can be seen as a unique substitute for continuous thinking about how to manage innovation, ignoring other valid approaches that are more focused on tacit KM.

Of course, the ability to build on new tacit knowledge is probably critical for success and a major challenge for management. The interplay between tacit and explicit is then a key issue to keep the game alive. But it is rarely recognized as such. It needs a dynamic approach to KM, but we are still struggling to design a static one!

5.5. Bibliography

Amin A., Cohendet P., 2004. *Architectures of Knowledge*. Oxford University Press.

Augier M., Shariq S., Vendelo M.T., 2001. "Understanding context: its emergence, transformation and role in tacit knowledge sharing", *Journal of Knowledge Management*, Vol. 5, No. 2, pp. 125–136.

Baird L., Henderson J.C., Watts S., 1997. "Learning from action: an analysis of the army lessons learned (CALL)", *Human Resource Management*, Vol. 36, No. 4; pp. 385–397.

Baumard P., 1999. *Tacit Knowledge in Organizations*. London: Sage.

Boughzala I., Ermine J.-L, 2004. *Management des connaissances en entreprise*. Paris: Hermès Science Publications.

Davenport T., Probst G. 2000. *Siemens. Knowledge Management Case Book*. Munich: Wiley & MCD Verlag.

Ermine J.-L., 2003. *La gestion des connaissances*. Paris: Hermès Science Publications.

Geisler E. 1999. "Harnessing the value of experience in the knowledge-driven firm", *Business Horizons*, May–June, pp. 18–26.

Grey D., 2002. http://www.voght.com/cgi-bin/pywiki?BoundaryObject.

Hansen M., Nohria N., Tierney T., 1999. "What is your strategy for managing knowledge?", *Harvard Business Review*, March–April, pp. 106–116.

Huber G., 1991. "Organizational learning: the contributing processes and the literatures", *Organization Science*, Vol. 9, No. 1, pp. 88–115.

Nonaka I. 1994. "A dynamic theory of organizational knowledge creation", *Organization Science*, Vol. 5, No. 1, pp. 14–37.

Nonaka I., Konno N., 1998. "The concept of 'ba': building a foundation for knowledge creation", *California Management Review*, Spring, pp. 40–54.

Nonaka I., Takeuchi H., 1995. *The Knowledge-creating Company*. Boston: Harvard Business School Press.

Polanyi M., 1958. *Personal Knowledge*. Chicago University Press.

Schrage M., 1995. *No More Teams*. New York: Currency.

Star S.L., 1989. "The structure of ill-structured solutions: boundary objects and heterogeneous distributed problem solving", in Gasser L. & Huhns M. (eds), *Distributed Artificial Intelligence*, pp. 37–54.

Wiig K., 2000. "Knowledge Management: an emerging discipline rooted in a long history", in Despres D. and Chauvel C., *Knowledge Horizons*. Boston: Butterworth-Heinemann, pp. 3–26.

Chapter 6

Approaches and Methods for Valuing Knowledge Management Performance

Over the past decade, Knowledge Management (KM) has developed rapidly in many organizations via various programs. These programs have been supported by specific organizational and technological structures (Knowledge Management Systems (KMS)) and aimed at optimizing the use of the knowledge of the employees of the firm. Objectives of these programs have been more precisely "*the generation, representation, storage, transfer, transformation, application, embedding and protecting of organizational knowledge*" [SCH, 02].

The existence of such programs demonstrates that firms trust the strategic value of knowledge. In fact such management practices introduce a new way of considering knowledge in a company: Knowledge is an asset that companies have to manage in an effective way in order to remain competitive [MAR, 94].

Nevertheless, even though several theories (such as the knowledge-based view, for example) give arguments that show a relationship between KM programs and organizational performance, this relationship is significantly more difficult to demonstrate empirically. Despite long practice in valuing the efficiency of material assets and organizational processes, this new "Knowledge Economy" leads firms to develop new methods in addition to valuing the efficiency of their intellectual and knowledge capital management. KM evaluation has become a strategic challenge for

---

Chapter written by Aurélie DUDEZERT.

managers because, as explained by J. Mouritsen, P. Bukh, H.T. Larsen and M.R. Johansen [MOU, 02], "what you can measure, you can manage, and what you can manage, you can measure".

This chapter reviews the state-of-the art on this specific subject. It aims to identify the different ways explored for valuing KM performance in organizations.

In the first part of this chapter, we focus on the empirical difficulties in valuing KM performance. The various definitions of KM and the different ways of characterizing performance explain the manifold approaches and methods to valuing KM performance in organizations. This analysis allows us to propose, in the second part of the chapter, a categorization of methods and approaches to valuing KM performance.

6.1. KM performance: the aims of an evaluation

6.1.1. *The knowledge-based view: a theoretical relation between KM and organizational performance*

From a theoretical viewpoint, the relationship between KM and performance has been considered in the theories of the resource and knowledge-based view (Penrose and Wernefelt, Prahalad and Hamel). In fact, these theories argue that to improve their performance, companies have to gain sustainable competitive advantages and that efficient KM provides firms with such advantages.

Nevertheless, such theories consider that to be efficient, KM programs have to manage "strategic" knowledge. KM can help companies improve their performance only if it concerns the management of "strategic" knowledge. However, the definition of "strategic" knowledge makes it very difficult to identify such knowledge. Actually, in these theories, knowledge has to be valuable, rare and isolated from imitation and substitution to be strategic [HOO, 03]. Unfortunately, such criteria are too general and abstract to be efficiently and operationally used in companies [MET, 02] [ARR, 01]. Therefore, it is difficult to demonstrate empirically the relationship between management of "strategic" knowledge and performance of a company.

Besides, if the abstract character of the concept of "strategic" knowledge were one of the reasons used to explain the difficulty of establishing empirically the performance of KM, we must also notice that in companies, the various definitions of KM and the different ways of characterizing performance make it difficult as well.

6.1.2. *What is performance for KM?*

The resource-based view (RBV) and knowledge-based view (KBV) theories consider the competitive performance of the firm: a company is efficient if it has a competitive advantage. Note that in the literature there are a lot of definitions of the performance of the firm. The competitive performance is one example among many others.

As observed by Reix [REI, 02], the performance of a firm is "multi-dimensional". The definition of the performance depends on the objectives of the company stakeholder who wants to value this performance. For instance, the shareholders and the employees will probably not have the same definition of the performance of the firm. Similarly, the definition of the KM performance depends on these different viewpoints.

In a study of the performance of information systems, Raymond [RAY, 02] has proposed the following three definitions of a firm's performance:

- the operational performance;
- the financial performance;
- the competitive performance.

When a firm's performance is defined as operational, methods and approaches to value the performance of information systems aim at identifying how information systems could improve the operations of a firm. From the financial performance viewpoint, methods and approaches aim to identify how information systems could improve the financial benefit and profit of the company. Finally, if the competitive performance definition is adopted, methods and approaches aim to identify how information systems could provide a competitive advantage. This categorization of the performance definitions is close to the categorization of the performance according to the firm's stakeholders. Indeed, the operational performance is related to the viewpoints of actors in the company who are directly involved in the operations of the firm: they adopt a "micro-organizational" viewpoint of performance. In contrast, the financial and competitive performances express a global or "macro-organizational" viewpoint of the company performance, which is more currently used by managers or strategists.

These two categories of definition of the firm performance also impact on the various approaches to the KM performance. However, the various definitions of KM and the multiplicity of forms of KM programs also explain the multiplicity of approaches and models for valuing KM performance.

6.1.3. *What is KM?*

It is common today to say that there are different forms of knowledge and various points of view on what is knowledge in organizations [NON, 91] [REI, 95] [BAU, 95] [ALA, 01] [CHA, 02]. Nevertheless, it is worth noting that there are still very different definitions of what KM in companies is or what it should be.

In the academic literature a quite consensual definition of knowledge management was given by Schultze and Leidner [SCH, 02]:

> Knowledge Management is the generation, representation, storage, transfer, transformation, application, embedding and protecting of organizational knowledge. Organizational memory, information sharing and collaborative work are closely related to knowledge management.

However, this definition establishes the very broad scope of KM and that very different types of management practice can be associated with KM. Indeed, in an operational way, such definitions make it difficult to distinguish between what KM is and what it is not.

KM can in fact have so different forms that Earl [EAR, 01] describes "Knowledge Management schools" and characterizes "systems, cartographic, process, commercial, organizational, spatial and strategic schools" of KM. That is the reason why we can reasonably wonder about the possibility of establishing one single framework for the evaluation of KM performance. Besides, Tuomi [TUO, 02] has noted that KM has its origins in four different disciplines that were relatively independent until the late 1990s. He characterizes these four disciplines of KM as "organizational information processing, business intelligence, organizational cognition, and organizational development". The first had its starting point in computer technology, the second in information services, the third in research on organizational innovation, learning and sensemaking, and the fourth in business strategy and human resource management. For this reason Tuomi considers that "Instead of one 'knowledge management', we therefore need several characterizations, which all remain somewhat ambiguous, overlapping, and depend on the point of time which we use". Consequently, KM programs in companies could be inspired by each of these four different disciplines and could be characterized by very different management practices with very different operational objectives. This variety of forms and origins of KM explains the variety and the multiplicity of approaches and models proposed for valuing KM performance in companies.

6.2. Method of research and inquiry

The multiplicity of points of view on KM performance makes it difficult to identify in an exhaustive way the methods used in companies to value the performance of KM.

In the first phase of our research work, we selected a sample of about 15 methods for valuing KM performance. These methods seem to us representative for highlighting the variety of KM evaluation approaches currently used in companies.

Very few academic papers present precise methods for valuing the performance of KM in organizations. That is the reason why, in order to identify these methods, we have used an inventory first developed by an Industrial Association, the European Industrial Research Management Association (EIRMA) and we have enhanced it. We have chosen to base our work on this first inventory because it had been developed by a consortium of industries. In that way we identified first the methods used by and/or well known in companies.

With the same objectives we completed this sample of methods by identifying methods whose developments had been supported by Public Administrations such as the European Commission (NIMMeasure, MAGIC, etc.) or by organizations that could easily transfer such methods to companies (APQC and Arthur Andersen, Knowledge Management Club in France, etc.).

In the second phase of our work we carried out a bibliographical study on "performance and knowledge management". Academic papers with theoretical assumptions that allow the highlighting of the dominant approaches, from empirical and theoretical points of view to characterizing KM performance in companies, have been identified.

We have identified two main approaches:

– a "macro-organizational" approach to KM evaluation;

– a "micro-organizational" approach to KM evaluation.

6.3. Macro-organizational approaches to valuing KM performance

The "macro-organizational" approaches to KM performance evaluation study the impact of KM on the global performance of the company (operational, financial and competitive), but they mainly focus on the financial and competitive performance of the company. These approaches are the oldest and the majority of works on KM performance adopt such a "macro-organizational" point of view.

6.3.1. *The competitive performance of KM*

Methods such as the *Balanced ScoreCard* [NOR, 96], the *Intangible Assets Monitor* [SVE, 98], the *Intellectual Capital Statement* [MOU,02] and *IcdVal* [BOU, 00] can be categorized in this approach. They are used to support the decision process. They provide information to improve the management of knowledge across the whole company in order to gain or improve competitive advantage for the company. Such methods value the performance of the KM process of the company.

The most well known methods for valuing the competitive performance of KM are the Balanced Scorecard and the Intangibles Assets Monitor.

The *Balanced Scorecard* was developed in 1996 after a study led by the Harvard Business School. It is a multidimensional measure system to help managers in their decision process. It is composed of indicators on the financial, the customer-relationship, the production and the organizational learning statements of the company. It aims to give to managers a global vision for managing the company efficiently. In fact, evaluation of the knowledge or organizational learning management is not the main objective of this method.

In contrast, the main objective of the *Intangible Assets Monitor (IAM)* is to value KM in a company. This method was developed by K-E. Sveiby and a consortium of industries. In this method, employees are considered as the main profitable resource of a company. Therefore, even though the frameworks of the IAM in human capital, customer capital, and organizational capital seem very closely related to those of the Balanced Scorecard, the goal is different. IAM aims to redefine the company as a knowledge company: "Sveiby attempts to redefine/re-evaluate the firm from the 'knowledge perspective'" [PET, 00]. Knowledge is the only useful resource for achieving a competitive performance and IAM helps to characterize the efficiency of the management of this knowledge.

Other methods are influenced by this vision of the KM performance. *ICdVal* considers knowledge as a resource that, combined with other resources in an efficient way, allows the competitive performance to be achieved. The *Intellectual Capital Statement* could also be put in this category. Indeed, this method aims at identifying the core knowledge of a company. With the help of the actors in a company this method analyses the customers' needs (for products or services sold by the company) and how knowledge associated with these needs is managed. In that way it allows identification of how knowledge of the company is able to fulfill the customers' needs and how the company can gain a competitive advantage.

Other methods adopt this competitive vision of KM performance. Nevertheless, they aim at focusing more specifically on the optimization of the management of knowledge rather on the optimization of the management of the whole company thanks to knowledge. Methods like *KMAT (Knowledge Management Assessment Tool)* developed by the American Productivity and Quality Center and Arthur Andersen [HIE, 96] [EIR, 99] and *KMM (Knowledge Maturity Model)* developed by the French Knowledge Management Club [ERM, 99] form part of such methods. Those methods have close similarities and put forward the hypothesis that there is a single way to have efficient KM for a company. KM of a company is compared with an ideal of efficient KM. If the company fulfils the criteria for having this efficient KM, the company has a competitive advantage and is efficient.

6.3.2. *The financial performance of KM*

Methods such as *Intellectual Capital of Strassmann* [STR, 96] [STR, 98] [STR, 99], *DOW's Knowledge Evaluation Method* [PET, 96] and *Tobin's Ratio* are often quoted as methods that value the financial performance of KM.

Often inspired by trademarks or license accounting methods, such methods, which demonstrate the impact of KM on company financial ratios, were developed a long time ago. Knowledge is considered as an asset that can be valued and accounted like a tangible asset. The difficulty is to make such an intangible asset financially valuable. KM is considered as efficient if it leads to financial revenues.

The *Intellectual Capital of Strassmann* defines knowledge as information. The corresponding method values the productivity of information in a company. So, according to Strassmann, it is necessary to value how the "asset" knowledge, formalized in "information", could have a benefit for the company in valuing KM performance. With the same point of view on KM performance, *DOW's Method* aims at valuing the knowledge formalized in "licenses". KM will be efficient if it allows creation of licenses that may be valuable on the financial market.

It is worth noticing that all these methods are based on different points of view on what is KM, what element (process, information, license, etc.) has to be valued to value KM performance efficiently and what are the final goals of KM programs. However, all these methods and approaches put forward the same hypothesis: KM is a global management practice that has an impact on the whole firm. That way, performance of KM must be valued at a global level of a company.

Other approaches and methods adopt another point of view. They consider that, even if KM could have a global performance, KM leads first to an operational performance. Such methods and approaches study how the implementation of KM could have an impact on the performance of operations of companies. Such approaches are more recent and we call them "micro-organizational" approaches to KM performance.

6.4. Micro-organizational approaches to valuing KM performance

This second category of approaches and methods brings together work inspired by research and methods used in Information Systems and work inspired by research and methods concerning the operational processes of companies (more particularly the quality approach).

6.4.1. *The process-based approach to the performance of KM*

The first approach of this category is the "process-based" approach to the performance of KM. To study KM performance, this approach aims at valuing the impact of the KM process on the operational processes and the activities of an entity (department, business unit, etc.). The goal is to optimize the management and performance of this entity. KM is considered as a specific operational process.

A lot of methods developed within this approach are implemented in R&D Departments. In fact, in this approach, KM has *a priori* the most important impact on entities where knowledge is crucial and knowledge is particularly crucial in R&D Departments.

Frank's [FRA, 03] and Jaime's [JAI, 04] research work could both be categorized within this process-based approach to KM performance. In both cases the objective is to characterize the impact of KM on improving processes and activities in R&D entities or in academic laboratories. In order to succeed, the work focuses mainly on the study of the impact of the support process for KM, i.e. the information sharing process, on operational processes and activities.

Evaluation methods like *MAGIC (Measuring and Accounting Intellectual Capital)* [WAG, 00] and *NIMMeasure* [ROT, 00] aim at valuing KM in such an operational way. They focus on R&D process development and identify the different phases where knowledge and KM has a particularly important impact.

MAGIC is a method whose development was supported by the European Commission. It aims at supporting the decision process and the management of knowledge in R&D operational processes. MAGIC identifies factors that lead the R&D Department to be efficient with knowledge and KM and compares this ideal situation with the actual situation. Knowledge belongs to the R&D organization and is not specifically related to the R&D process.

NIMMeasure's development was also supported by the European Commission. It aims to optimize the management and development of knowledge in R&D processes. In contrast to MAGIC, this method considers knowledge as intimately related to the research process. KM is a specific process, but it is closely linked to the R&D process.

6.4.2. *The systemic approach to the performance of KM*

A second "micro-organizational" approach to KM performance focuses more specifically on the study of the performance of technological and organizational systems related to KM. Such research work and methods are inspired by research work on information technology (IT) and information systems (IS) performance. A lot of KM programs are today supported by information technology and information systems. That is the reason why some of the KM performance evaluation methods are inspired by works on IT and IS performance.

Such research work more specially studies the performance of the KM systems (KMS). The KMS is defined as a technological and organizational structure that supports KM.

Such approaches are still developing. Nevertheless, research such as Lancini's PhD on the identification of successful factors in KMS adoption in an organization [LAN, 01] [LAN, 03] can be put in this category. The performance of KM can be defined by analyzing the impact of the implementation of the KMS on the operational activity of the department. Lancini addresses this question by putting forward the hypothesis that the KMS is performing according to the level of use of the KMS in the department where it is implemented.

The *eSmac Method* [DUD, 03] is inspired by such IS approaches to KM performance. This method identifies the performance of KM by analyzing how the implementation of the KMS in a department allows the strategic objectives of the Department to be achieved. eSmac aims at identifying the value added by the KMS to the knowledge of the department and how this knowledge can impact the operational process of the department.

All these "micro-organizational" approaches are still developing. However, they are supported or developed by organizations intimately linked to companies (European Commission, French CIFRE PhD, etc.). They fulfill operational needs that are more and more explicitly expressed by companies, i.e. to identify precisely the concrete impact of KM on the operational processes of the company.

6.5. Conclusion

This chapter discusses the state-of-the-art of the different approaches to evaluating KM performance in companies. On a first reading one may be surprised by the variety of approaches and methods gathered. However, as we have explained in the first part of this chapter, this variety may not be the result of a bad definition of what KM is or what the performance of KM is. The variety of points of view on KM and performance lead naturally to the development of different approaches to evaluating the performance of KM.

Consequently, a manager can legitimately wonder about the best way to value KM in a company. Following our discussion of the state of the art, two main approaches appear to be possible:

– to adopt a specific definition of KM and performance and choose to build a specific approach to evaluating KM performance;

– to adopt a global vision of KM performance and value KM in a micro-organizational and a macro-organizational way.

In a knowledge economy, the second suggestion seems to be the most appropriate.

6.6. References

[ALA, 01] Alavi M., Leidner D.E, "Knowledge Management and Knowledge Management Systems: conceptual foundations and research issues", *MIS Quaterly*, vol. 25, no. 1, pp. 107–136, March 2001

[ARR, 01] Arrègle J-L., Quélin B., "L'approche fondée sur les ressources", in *Stratégies Actualité et Futurs de la Recherche*, edited by A-C. Martinet and R-A. Thiétart, Vuibert, FNEGE, Paris, 2001, pp 273–289

[BAU, 96] Baumard P., *Organisations déconcertées, la gestion stratégique des connaissances*, Masson, Paris, 1996

[BOU, 00] Bounfour A., "La valeur dynamique du capital immatériel", *Revue Française de Gestion*, no. 130 September–October, 2000

[CHA, 02] Charlot J.-M., Lancini A., "De la connaissance aux systèmes d'information supports", in *Faire de la Recherche en Systèmes d'Information*, (Chap. 8, pp 139–159), edited by F. Rowe, Vuibert, FNEGE, 2002

[DUD, 03] Dudezert A. *La valeur des connaissances en entreprise: recherche sur la conception de méthodes opératoires d'évaluation des connaissances en organisation*, PhD, Ecole Centrale Paris, 2003

[EIR, 99] EIRMA, *La gestion du savoir dans l'entreprise*, European Industrial Research Management Association, Report, 1999

[ERM, 99] Ermine J-L., "Evaluer la maturité cognitive d'une organisation", *BASES* no. 156, 1999

[FRA, 03] Frank C., *Développement de modèles spécifiques de gestion des connaissances dans les services de recherche*, PhD Institut National Polytechnique de Grenoble, 2003

[HIE, 96] Hiebler R., "Benchmarking: Knowledge Management", *Strategy & Leadership*, Vol. 24 no. 2, pp. 22–29, March–April 1996

[HOO, 2003] Hoopes D.G., Madsen T.L., Walker G., "Why is there a resource-based view? Toward a theory of competitive heterogeneity?", *Strategic Management Journal*, 2003, Vol. 24, pp. 889–902

[JAI, 04] Jaime A., Gardoni M., Mosca J., Vinck D., "La démarche qualité, cadre de la capitalistion des connaissances dans les organismes de recherche", Colloque IPI, Autrans, France, 2004

[KAP, 96] Kaplan R. S., Norton D. P., "Linking the balanced scorecard to strategy", *California Management Review*, Vol. 39, no. 1, Fall 1996

[LAN, 01] Lancini A., *Les déterminants de l'adoption d'un système de Gestion des Connaissances: contribution à l'étude du succès de la technologie Lotus Notes dans une société mutuelle d'assurance*, PhD, Université des Sciences Sociales Toulouse 1, 2001

[LAN, 03] Lancini A., "Les déterminants du succès des Systèmes de Gestion des Connaissances (SGC): étude de cas d'une mutuelle d'assurances", *Proceedings of the Colloquium of the Association Information and Management, 22 and 23 May 2003,* Grenoble, 2003

[MAR, 94] March J.G., "L'avenir de la gestion vu par James G. March", *Revue Française de Gestion*, September–October 1994, no. 100, pp. 22–27

[MET, 02] Metais E., "Vers la notion de polyvalence stratégique", *Revue Française de Gestion*, no. 138, April–June, 2002

[MOU, 02] Mouritsen J., Bukh P., Larsen H., Johansen M., "Developing and managing knowledge through intellectual capital statements", *Journal of Intellectual Capital*, Vol. 3, no. 1, pp. 10–29, 2002

[NON, 91] Nonaka I., "The knowledge creating company", *Harvard Business Review*, November–December 1991, pp. 96–104, 1991

[PET, 96] Petrash G., "Dow's journey to a knowledge value management culture", *European Management Journal*, Vol. 14, no. 4, pp. 365–373, August 1996

[PET, 00] Petty R., Guthrie J., "Intellectual capital literature review measurement, reporting and management", *Journal of Intellectual Capital*, Vol. 1 no. 2, pp. 155–176, 2000

[RAY, 02] Raymond L., "L'impact des Systèmes d'Information sur la performance de l'entreprise", in *Faire de la Recherche en Systèmes d'Information*, edited by F. Rowe, VUIBERT, Paris, 2002, pp. 301–320

[REI, 95] Reix R., "Savoir tacite et savoir formalisé dans l'entreprise", *Revue Française de Gestion*, no. 105, September–October, 1995, pp. 17–28

[REI, 02] Reix R., "Systèmes d'Information et performance de l'entreprise étendue", in *Faire de la Recherche en Systèmes d'Information*, edited by F. Rowe, Vuibert, Paris, 2002, pp. 333–350

[ROT, 00] Roth N., Prieto J., Dvir R., "New-use and Innovation Management and Measurement Methodology for R&D", *6th International Conference on Concurrent Enterprising Proceedings*, Toulouse, 2002

[SCH, 02] Schultze U., Leidner D.E., "Studying Knowledge Management in Information Systems Research: discourses and theoretical assumptions", *MIS Quaterly*, vol. 26, no. 3, pp. 213–242, September 2002

[STR, 96] Strassmann P., *The Value of Computers, Information and Knowledge*, Strassmann Inc, www.strassmann.com/pubs/cik/cik-value.shtml, 1996

[STR, 98] Strassmann P., "Taking the measure of knowledge assets", *Computerworld*, Vol. 32, no. 14, p 74, 6 April 1998

[STR, 99] Strassmann P., "Calculating Knowledge Capital", *Knowledge Management Magazine* (October), http://files.strassmann.com/pubs/km/1999-10.pp, 1999

[SVE, 98] Sveiby K-E., "Intellectual Capital: thinking ahead", *Australian CPA*, vol. 68, no. 5, pp. 18–22, June 1998

[TUO, 02] Tuomi I., "The future of Knowledge Management", *Lifelong Learning in Europe (LlinE)*, Vol. VII, issue 2, pp. 69–79, 2002

[WAG, 00] Wagner K., Hauss I., "Evaluation and measurement of R&D knowledge in the engineering sector", in *Challenges of Information Technology Management in the 21st Century, 2000 Information Resources Management Association International Conference*, Hershey, PA, USA, 2000

Name	BSC Balanced Scorecard	IAM Intangible Assets Monitor	ICdVal Valeur Dynamique du Capital Immatériel (Dynamic Value of Intangible Capital)	IC Statement	KMAT Knowledge Management Assessment Tool	KMM Knowledge Maturity Model
Evaluation objectives	Financial promotion of knowledge To support the management of knowledge in the global management of the company	Financial promotion of knowledge To support the management of knowledge in the global management of the company	Financial promotion of knowledge	To support the management of knowledge in the global management of the company	To optimize the management of knowledge	To optimize the management of knowledge
Stakeholders concerned by this evaluation	External audience/ Internal audience	External audience/ Internal audience	External audience	Internal audience	External audience/ Internal audience	Internal audience
Evaluation techniques	Quantitative	Quantitative/ Qualitative	Quantitative	Qualitative	Qualitative	Qualitative
Evaluation principles	Indicators of customer, human and organizational capital value	Indicators of customer, human and organizational capital value = redefine the firm in terms of the knowledge perspective	Performance indicators of resources, competences, processes and outputs and definition of a global indicator to identify the dynamic value of the intellectual capital	Definition of a "knowledge narrative" (core and crucial knowledge and KM processes) and comparison between this knowledge narrative and the actual situation of the company	Comparison of the actual situation with an ideal of KM Performance	Comparison of the actual situation with an ideal of KM Performance
Authors	Norton and Kaplan (Norton and Kaplan, 1996)	K.-E. Sveiby and Skandia (Sveiby, 1997)	A.Bounfour (Bounfour, 2000)	Scandinavian team (Mouritsen, 2002)	Arthur Andersen and American Productivity and Quality Center (APQC) (Hiebler, 1996; (EIRMA, 1999)	J.-L. Ermine and the Knowledge Management Club (Ermine, 1999)

Table 6.1: *Macro-organizational approach: detail of the methods to value the competitive performance of KM*

Name	Knowledge capital	The DOW method
Evaluation objectives	Financial promotion of knowledge	Financial promotion of knowledge
Stakeholders concerned by this evaluation	External audience	External audience
Evaluation techniques	Quantitative	Quantitative
Evaluation principles	Intellectual capital = Information productivity	Knowledge value = License revenue
Authors	P. Strassmann (Strassmann, 1996) (Strassmann, 1998) (Strassmann, 1999)	Dow Chemical Company (Petrash, 1996) (EIRMA, 1999)

Table 6.2: *Macro-organizational approach: detail of the methods for valuing the financial performance of KM*

	Process-based approach		***Systemic (technological and organizational) approach***
Name	MAGIC Measuring and Accounting Intellectual Capital	NIMMeasure	eSmac
Evaluation objectives	To support the management of knowledge in the R&D processes	To support the management of knowledge in the R&D processes	To support the management of Knowledge Management Systems (KMS)
Stakeholders concerned by this evaluation	Internal audience	Internal audience	Internal audience
Evaluation techniques	Quantitative/Qualitative	Quantitative	Qualitative
Evaluation principles	Factors of success determined by an analysis of the R&D strategy and the market needs and comparison with the actual situation	To value the impact of knowledge in the R&D process	Identification of the value-added of the KMS on the knowledge of the entity where the KMS is implemented
Authors	European Project MAGIC (Wagner, 2000)	European Project NIMCube (Roth, 2000)	(Dudezert, 2003)

Table 6.3: *"Micro-organizational" approaches: detail of the methods*

Chapter 7

ICIS for Knowledge Management: The Case of the Extended Enterprise

7.1. Introduction

To respond to an economic and competitive environment that is becoming more and more complex, as well as to the continual internationalization and the globalization of markets, firms opt more frequently for alliances and partnerships, developing new forms of organization based on inter-company relations. The emergence of the concept of the extended enterprise is the result of the hope of some companies that they can form alliances with others, after careful selection, and set up long-term and valuable exchange relationships. Inter-company Co-operative Information Systems (ICIS) appeared to support these relationships of co-operation, namely communication, co-ordination and collective problem-solving and to manage knowledge, which proves to be a delicate task when the large of participants is considered.

In this chapter, we put forward a contribution to the design of ICIS that is based on information systems engineering [CAU 2001] and knowledge engineering [CHA 2000] in the context of inter-company co-operative activities. The goal is to guide and to facilitate the design of software supporting inter-company co-operation and Knowledge Management (KM).

We begin by defining inter-company co-operation and by presenting the current situation of the extended enterprise as a particular form of this kind of co-operation,

---

Chapter written by Imed BOUGHZALA.

and discussing the reasons for which KM issues are particularly delicate. Then after we have justified the different levels that seem necessary to us to achieve the intended aims, we introduce different models through a methodological framework, aiming to design an information system for the extended enterprise that ensures KM within the setting of inter-company co-operation.

7.2. Concepts and definitions

7.2.1. *Inter-company co-operation and the extended enterprise*

Inter-company co-operation describes a situation where two or more partner agents, under the aegis of contracts, share their resources and their complementary means, for problem-solving in order to accomplish one or several common activities. These agents communicate in order to co-ordinate their tasks.

Among the different forms of co-operation, let us consider the reticular forms of companies such as the networked enterprise or V-Firm – V for vertical co-operation between donors and subcontractors for example, or the network of enterprises or H-Firm, covering horizontal co-operation between subcontractors [ALB 96]. The notion of the extended enterprise that includes both these reticular forms we choose to call the D-Firm, D referring to diagonal co-operation because it is both horizontal and vertical [BOU 01]. This concept of the extended enterprise is a form of organization including all partners, suppliers, donors, subcontractors, competitors, etc. It is not a network of enterprises or a networked enterprise but a fully fledged organization.

We consider the extended enterprise as a set of partner agents sharing resources and complementary competencies, similar or not, that are co-operating in order to reach shared objectives. The structure of this type of company is based upon the existence of a human co-operator agent (i.e. a consulting company) who ensures co-operation between partners (business agents). This agent represents the mediator in this type of enterprise while allowing the establishment of a global setting of co-operation. This neutral agent, in that it does not participate in the effective co-operation, is also a catalyst for or a regulator of the extended enterprise. A co-operator agent can manage one or several extended enterprises and a business agent can belong to several extended enterprises managed by different co-operator agents.

7.2.2. *ICIS and KM*

Several definitions have been given for Co-operative Information Systems (CIS). Distributed Artificial Intelligence community (DAI) defines a CIS as a combination of *"computational" agents who continuously share objectives with other information systems, both human agents and organizations within their operational environment* [HER 01]. In the distributed database approach, *a CIS integrates distributed information sources – databases or knowledge-based systems – which can use knowledge representations and heterogeneous data* [BRO 92]. In the domain of the CSCW, *a CIS is more or less a set of autonomous pre-existing components that usually work in a synergistic manner by exchanging information and expertise and by coordinating their activities* [BOU 97]. On the basis the principal ideas in these definitions, we describe the ICIS (a specific CIS in the inter-company context) as an information system that assists the collective work between different agent types (individuals, units, groups, enterprises, etc.). Such an IS contains intrinsic specificities related to the potentially heterogeneous character of the information sources, processes and knowledge relationships, and is also linked to the independence that agents wish to preserve. This type of system encourages, on the one hand, the nomadic capacities and the agent's mobility with the same functionalities and security levels that exist within an enterprise and, on the other hand and in particular, the integration of enterprises. While it is for information systems architects to study and conceive systems that offer the same functionalities as classic information systems offer, such systems should support collaborative work. This is a new domain that borrows from various existing domains, which include information systems, distributed artificial intelligence, distributed databases, management sciences, sociology, and education sciences.

One of the keys to a successful extended enterprise is its information system (as in the case of all enterprises), which is referred to in this context as the ICIS. This system, preferably managed by the co-operator agent, brings together actor groups, either restricted or extended, belonging to legally independent organizations, each taking one of three forms of information system according to the role each has to play in inter-company co-operation processes [BOU 02]:

– a communication IS, which facilitates the communicative interaction between companies;

– a co-ordination IS, which facilitates the co-ordination of inter-company activities;

– a collective problem-solving IS, which facilitates the collaboration between enterprises working, in a simultaneous way, on common tasks.

These three forms of IS allow three types of inter-company integration to be established:

– an integration through data when companies are brought together only to exchange data;

– an integration through processes when companies are brought together to co-ordinate their processes (for example, a process starts in one enterprise and ends in another or is completely externalized within one or the other);

– an integration through knowledge when partners share knowledge to accomplish a common task or to create and enrich each other's knowledge.

Technically, we can speak about an integration through applications to cover all three types of inter-company integration levels. The use of web services technologies can be a good choice in such a case.

As explained convincingly by many authors ([STE 93]; [ERM 00]; [GRU 96]), the pressure of technical and organizational innovations, new methods of management (management by project) and voluntary or enforced staff mobility (transfers, redundancies, retirements) explain why KM has become such a major preoccupation (stake) in numerous organizations. Knowledge systems management (management by knowledge) plays an enormous role within large organizations, and even when firms are members of an extended enterprise. Even when firms are focused on their own know-how, they are still extremely dependent on their partners concerning adjustment of their production and concerning their innovation processes. These companies must always be up-to-date with their partners' evolution so as to be better able to anticipate new needs and future constraints. In inter-company co-operation, we can distinguish three kinds of knowledge [TAT 03]:

– piloting knowledge representing the goal of co-operation;

– functional knowledge representing the object of co-operation;

– operational knowledge representing the implementation of co-operation.

These kinds of knowledge can be private and confidential, private but shareable or public.

Indeed, the ICIS contains information and knowledge with a syntax and semantics that are shared and understood by all members of the extended enterprise. It represents in some ways the *extended enterprise memory* (EEM) [DIE 00], which allows memorization of shared or generated information and knowledge of the inter-company co-operation in the extended enterprise, a beneficial quality that is an advantage to all members. It must enhance the advantages of Internet and Web technologies (extranet, e-community, web services, etc.).

The co-operator agent depends on this ICIS to ensure the management of information, of processes and of knowledge in the extended enterprise, but also beyond that, to manage for example, social conflicts that can emerge between partners.

An adapted ICIS can be regarded as a capital stake for the extended enterprise. It is this that led us to propose a methodology, called MeDICIS (*Methodology for Designing Interenterprise Co-operative Information System*), as a design for this type of IS. Our proposal is intended to control the complexity of the co-operation processes and the difficulty of building an effective ICIS, while also providing help for the software specification for the ICIS to enable KM within the extended enterprise.

7.3. MeDICIS

To respond to the need for a design for an ICIS, we propose MeDICIS, a methodological framework that aims to guide the design of inter-agent co-operation mechanisms. The goal is first to analyze, then to model co-operative processes according to three levels of depth (communication, coordination and collective problem-solving) and then to specify an ICIS adapted to the needs and the constraints of co-operation in order to be able, finally, to manage the knowledge used and generated by this co-operation through the ICIS.

Concerned with IS and knowledge engineering, the models that we present here relate to the two levels of modeling that MeDICIS identifies:

– a macro level corresponding to a generic modeling of co-operation that is interested in the business setting of inter-company co-operation (e-business);

– a micro level, in which we model, on the one hand, the course of co-operation according to its three levels of depth – communication, co-ordination and collective problem solving (CPS) – and, on the other hand, features of agents (representing their enterprises) implied in co-operation. It should be noted that the modeling of a level of depth implies inclusion of the levels that precede it; for example, modeling CPS implies a priori the modeling of communication and co-ordination.

The models suggested are as follows:

– At the *macro* level: the *business model* gives the general setting of the business, its structure and its dynamics, the business contract, the business participants, their resources, services and products. It makes it possible to give a general view of the distribution of tasks between partners (business agents).

– At the *micro* level:

- The *agent model* describes the different types of agents, their features and their relationships; for example, an agent can be artificial, actor, group of actors or organizational entity (business unit), with its specific knowledge, its competencies, its qualifications, its roles, etc. This model can be used, for example, to define a personnel directory of a company.

- The *co-operation model*, which allows identification and understanding of the processes generated by the business.

- The *communication model*, which provides a specification for the communication put, or to be put, in place: the interlocutors, the exchanged messages, the context, the channel, etc.

- The co-ordination model, which determines the course (progress) of the co-ordination of activities between agents, their roles, the necessary resources, entry and exit flows, etc.

- The Collective Problem-Solving (CPS) model, which describes the realization of tasks in solving a particular problem.

These models and their organization are presented in Figure 7.1.

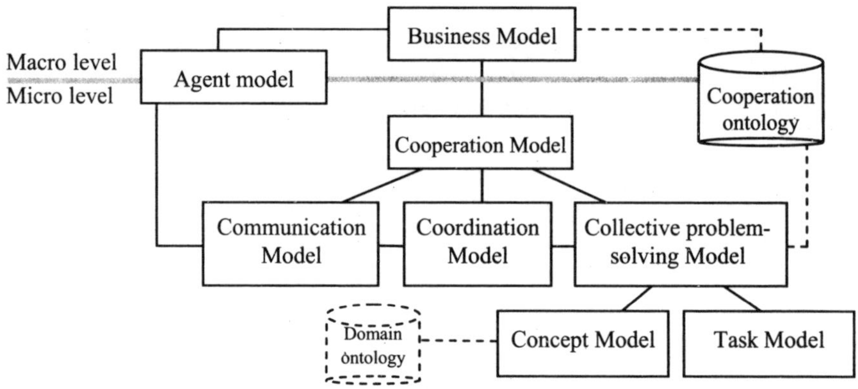

Figure 7.1. *Levels and models of MeDICIS*

The possible existence of a co-operation ontology, shared on the different levels, should be noted here. This co-operation ontology is proposed so as to overcome differences in language, variety of representations and the heterogeneity of ontologies peculiar to each partner (vocabularies, categorization, etc.). It thus allows communication between members of the extended enterprise, even though they work in different contexts, different professions, with different needs and different view points [OLE 98]. It provides referential, conceptual and terminological principles concerning the models of MeDICIS. These models, discussed below, draw on this ontology to describe their objects (product, service, agent, resource, activity, message, expertise, knowledge, etc.). Moreover, and beyond the aspects that interest us directly here, the ontology encourages reuse, reliability and specification of an ICIS, in particular, of an extended enterprise memory while lowering the cost of realization.

Levels of MeDICIS	Components of MeDECIS	UML elements	MASK elements	Other sources of inspiration
Macro	Business model	Use case diagrams Class diagram		UMM, UEML, UDDI
Micro	Co-operation model	Use case diagrams Class diagram		UEML
	Agent model	Classes diagram		CommonKADS [SCH 99]
	Communication model	Class diagram Sequence diagram		MAS (Multi-Agent Systems)
	Coordination model			SADT, UML (Activity diagram), MASK, Workflow: IDEF3
	CPS model		Concept model Task model	
Tool specification		Use case diagrams Component diagram Deployment diagram		Others

Table 7.1. *MeDICIS components*

Our proposal is based on the conceptual and graphic conventions of modeling languages such as UML (Unified Modeling Language) and elements of the method of knowledge systems analysis, MASK [ERM 01] (cf. Table 7.1). We are not proposing here a "new formalism" but one that is integrated in already tested steps of IS design and of knowledge engineering [RUM 99]. Table 7.1 shows, for each model, elements taken from UML and MASK as well as those borrowed from other sources such as UMM [UMM 01] or from the XML (eXtensible MarkUp Language) diagram of UDDI[1].

In what follows, we will endeavor to describe only the models most characteristic of our proposal. Then, we will present the life cycle of MeDICIS.

1 Universal Description, Discovery and Integration is an universal business register intended to improve exchanges between firms on the Internet (http://www.uddi.org).

7.4. Models for ICIS design

7.4.1. *The business model*

The business model describes the general setting of the business in the extended enterprise. This model allows partners to participate in B2B electronic commerce practices or "e-business" on the Internet. It is an exchangeable model that allows interoperation of heterogeneous systems.

The design of the model rests on the use case and class diagrams of UML. The class diagram (see Figure 7.2) allows us to identify and to describe elements of this business setting (business sector, business process domain, events launching the business, etc.). It puts forward business agents (partners in the extended enterprise), the contractual context that binds them through the co-operator agent, business resources, products and services that they perform.

This model also makes it possible to count all shared business documents and to know the distribution of activities and tasks assigned to each partner according to the formalized protocols, as well as co-operative tasks involving several agents. The co-operator agent only appears in this model since he ensures the business setting and doesn't participate in the actual co-operation.

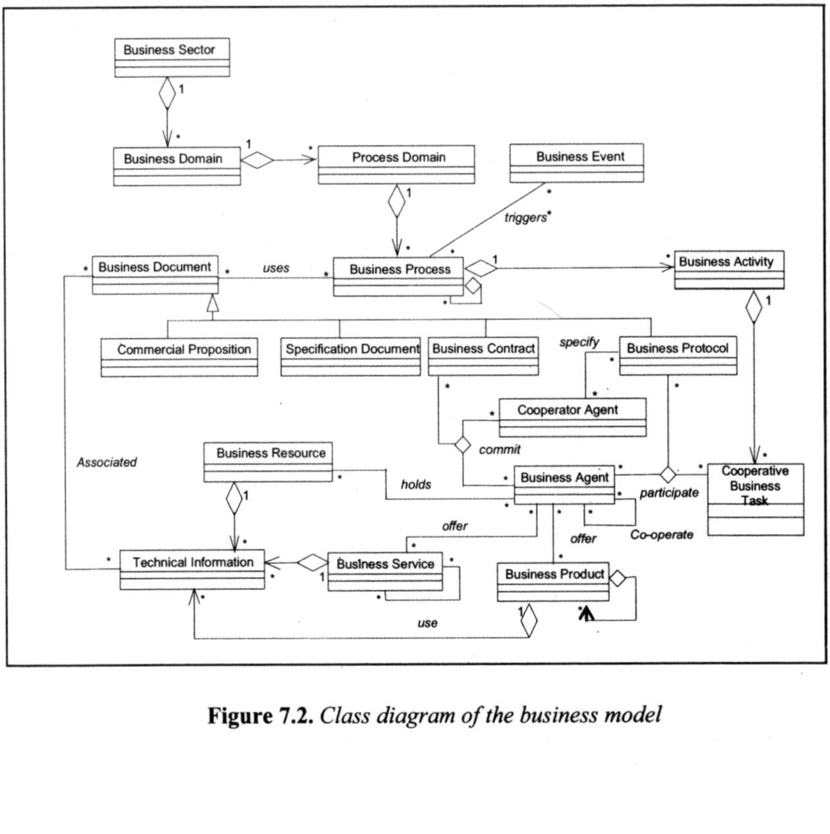

Figure 7.2. *Class diagram of the business model*

We note that this diagram is a major asset in helping the co-operator agent meet the customer requirements and for the partners to recognize their commercial identity, their resources, their products and/or services and to have a global vision of the business.

7.4.2. *The co-operation model*

This model makes possible to refine the business model so that the course of co-operation between agents (as an operator: machine, actor, business unit) belonging to different business agents (partner companies) can be described activity by activity. Every business activity is turned, in this case, into a process of co-operation.

The use cases feature allows us to represent the dynamics of processes in the modeled co-operation domain and the class diagram allows to identify and to describe objects of co-operation. We were also inspired by the specifications of UEML [VER 96] to adapt it to the specific inter-company co-operation processes.

Through the class diagram of this model (Figure 7.3), we distinguish between objects of co-operation: they are involved in either entry to or exit from the co-operation process. For example, we can imagine as an entry object of co-operation the specifications formulated by the customer, and as an exit object the product requested. The result of this model is the division of the process into co-operative activities and tasks mobilized by specific roles (qualifications and precise expertise) as well as the implied constraints. We distinguish between five generic classes of resources, from which all other types can be derived, as suggested in CIMOSA (Computer Integrated Manufacturing Open System Architecture – [SCH 99]; [VER 96]):

– computer applications;

– humans;

– machines;

– information (market study, various documents, data bases, etc.), which we have added;

– knowledge (methods, techniques, processes, "best practices", knowledge book, knowledge basis, etc.).

This model acts as a tool that assists monitoring and management.

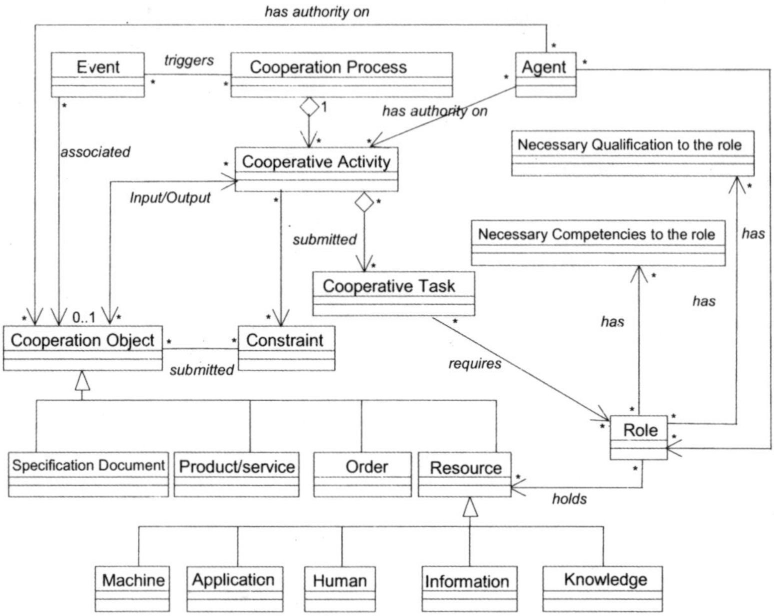

Figure 7.3. *Class diagram of the co-operation model*

7.4.3. *The agent model*

This model allows us to describe features of agents and their relationships according to UML class diagrams (Figure 7.4). It allows us to visualize the agent's type. An agent defines an element in a structure and provides control and responsibility for the identified co-operative activities and the objects of co-operation. An agent can be composed of (or be responsible for) several agents. It can be an artificial agent (machine or computer, application or program, automaton), an actor (individual), a group of actors (internal or inter-company work group, team project), or an organizational entity (business unit, division, direction, department, etc.). An agent is characterized by its geographical localization, its expertise (professional know-how), its knowledge (theoretical knowledge) and its qualifications (diplomas or certification). According to these characteristics, some or several roles are assigned to it according to a specific task.

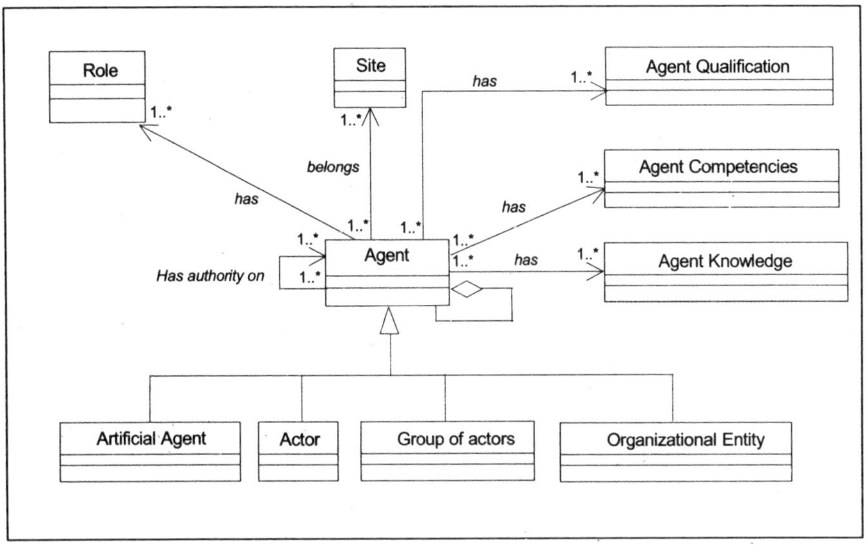

Figure 7.4. *Class diagram of the agent model*

7.4.4. *The communication model*

This model allows us to identify, understand and put in place agreed approaches to communication. This model assists agents to keep a record of their exchanges subject to their memory limitations.

We associate with this model a class diagram (Figure 7.5) that allows us to represent the static "spatial" structure of communication; the sequence diagram will help us to describe the chronological aspect of communication and dialogue scenarios between agents.

In this diagram we distinguish four types of agent that can be involved in a communication situation: an issuing agent, a receiving agent, a supervisor agent (a control function) and an observer agent (for information). The code guarantees the understanding of messages exchanged by the communicating agents. The channel is the means of transmission the method of message circulation (for example, AVN – added value networks – or the Internet). Localization is the space used for storing the exchanged messages.

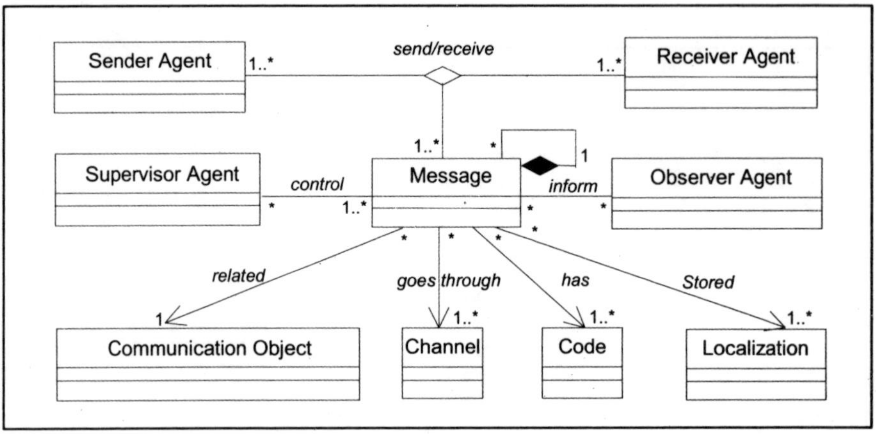

Figure 7.5. *Class diagram of the communication model*

7.4.5. *The co-ordination model*

This model was inspired by the activity model of MASK and therefore by the SADT actigram, by the UML activity diagram and workflow representations, especially that of IDFE3 ([MAY 95], [WFMC 99]).

It makes it possible to identify, understand and put in place mastered agreed co-ordination situations. It especially serves to facilitate the co-ordination of interdependencies between agents' partners. It ensures co-ordination of operations between partners all along the chain of value for a given product or service. As flows of control cannot be mixed with material flows, it helps to master the circulation of materialized flows (data, material, staff, monetary, etc.) or of work/control flows; but does not allow the two types of flux to be mixed during a co-ordination process. One of the major enrichments of this model is the use of several types of routing. To the three types of routing (sequential, parallel, conditional) of activity diagrams, three types inspired by workflows (conjunction – And-join– disjunction – OR-join and XOR-join – and Loop) are added, as illustrated in Table 7.2.

Sequential	⟶
Parallel	┬ (split into parallel arrows)
Conditional	◇
Conjunction (And-join)	⊕
Disjunction (OR-join, XOR-join)	○ **O**
Loop	⊕ (in diamond)

Table 7.2. *Routing of activities*

The chart of a co-ordination model is as shown in Figure 7.6:

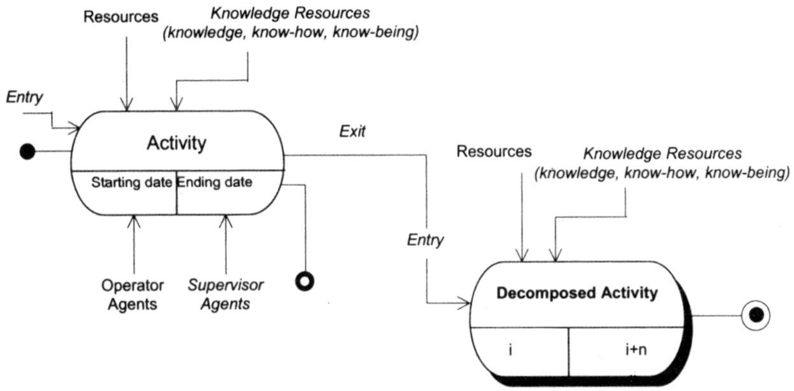

Figure 7.6. *The co-ordination model*

In this model we distinguish between operator agents and supervisor agents, and, as in MASK, we distinguish between ordinary resources and knowledge resources (knowledge, know-how and we add know-being). The temporal references are marked on the activity (date beginning, date end). Knowing the length of an activity is an important parameter for project management (Pert, Gantt) and a necessary factor in succeeding at a better co-ordination of inter-company activity.

7.4.6. *The CPS model*

This model is made up of the two models from MASK, concept and task, [ERM 01] for the modeling of collective problem-solving in an inter-company co-operation.

The concept model represents the static aspect of knowledge including the conceptual structuring (ontology of domain or catalog of concepts) of an expert, a person accustomed to work in a precise domain. "This structuring is given in the form of a classification of concepts, of objects of the domain. The classification is a basic natural cognitive activity for a human being. All knowledge is characterized by the possibility of giving a detailed taxonomy of the domain, built up from experience" [ERM 01].

The task model represents the dynamic knowledge and so describes ways in which to achieve a task or to solve a problem. It "includes the description of the scheduling of tasks to be accomplished in order to have a decomposition put into a tree hierarchy, which recursively refines tasks from a higher level to more detailed sub-tasks until there are tasks that can be considered as terminals" [ERM 01]. The type of a given task expresses the control exercised on the sub-tasks that decompose it.

7.5. Discussion

With reference to levels of partners' integration in an extended enterprise mentioned above, we raise a matrix of correspondence that shows the relation between levels of co-operation depth, integration of enterprises, proposed models and tools.

Co-operation depth Levels	**Inter-company integration Levels**	**Proposed models**	**Technologies (tools)**
Communication	Integration by data	Communication model	Email, Chat, EDI[2], Visio conference, white board, etc.
Coordination	Integration by processes	Coordination model	Library, Calndar, Shared Agenda, Workflow, Project Plan, etc.
Collective problem-solving	Integration by knowledge	CPS model	Shared Applications, Design Rational, GDSS, Competency Mapping, Poll, Ontology, etc.

Table 7.3. *Correspondence matrix*

2 Electronic Data Interchange.

7.6. MeDICIS life cycle

MeDICIS was worked out with three precise objectives:

– to analyze and then to model interagent co-operative processes in the case of implementing new inter-company co-operation or, in the case of existing co-operations, auditing. The goal here is to organize better co-operations and optimize them;

– to specify ICISs, adapted to the needs and constraints of co-operations, such as an extranet, collaborative workspaces/workplaces, e-communities (communities of practice, of interest, of learning ...), e-procurement tools, dynamic portals, and an e-market place;

– to manage the knowledge used and generated by these co-operations through an ICIS.

The life cycle of MeDICIS proposed breaks up into five phases:

– *Framing phase*: consists of defining the purpose and the objectives of the co-operation, to study its feasibility and its added value and to choose suitable partners.

– *Analysis phase*: consists of a general description of the co-operation and its actors by using business and agent models. It is a question of setting up of the storyboard of the co-operation in terms of business protocols, planning, contracts, etc.

– *Scenario phase*: has as its goal a description of the different co-operative activities as well as the actors involved in these activities (their roles) by using the co-operation model for each activity.

– *Modeling phase*: has as its goal the determination of the level of depth of each co-operative activity according to the more adapted type of integration (by data, by process, by knowledge). The choice of model (communication, coordination and CCP) is made according to the corresponding depth levels. The concept of domain ontology can prove to be essential when it provides integration by knowledge; in this case we will call upon knowledge mapping techniques developed by the "*club gestion des connaissances*" [AUB 03]. This phase can reach its end if it produces a simple co-operation audit and the audit report is delivered with all the documents and models produced since the beginning of the life cycle of the method.

– *Specification phase*: consists of the specification of an ICIS based on collaborative tools. The choice of tools is made according to the analysis carried out in the preceding phase. These tools can be designed specifically on the basis of "measurement" (customized) or packaged through existing tools on the market. This phase proceeds exactly as for a traditional IS design project.

7.7. Conclusion

In this work, we tried to tackle the specification of ICIS, related to the potentially heterogeneous character of sources of information, of processes and of knowledge and with the independence of agents, we proposed a contribution to the design of these IS, through models taking of account constraints of communication, co-ordination and collective problem-solving specific to this context. Our goal not being to still propose "another model", we privileged the re-use of existing formalisms and models, in order to integrate largely tested methodological frameworks.

The evaluation was carried out through several experimentations which we will not develop in detail in this chapter:

1. In the setting of the textile field, through a project of specification of a prototype tool for inter partner electronic data interchange of WebEDI type implementing the properties of the communication model in the setting of a textile portal.

2. In the setting of the automobile field, through an audit of co-operation to analyze a business process in design of a new model of top-of-the-range car.

3. In the setting of the industrial design field, through a knowledge cartography of a community of practice of designers and the setting up of a multi-project management tool in design in the setting of a E-business platform called Ecopics at INT (Institut National des Télécoms), implementing the properties of the agent, communication and coordinations models.

These evaluations made it possible to better target certain models, then to validate them in a sequence carried out to a methodological framework.

Partner companies share information and knowledge, MeDICIS proposes to manage knowledge relating to the co-operation in question. Compared to the existing methods and frameworks, this methodology presents the particular interest to be specific to the modeling of inter-company co-operation, while re-using validated formalisms or methodological elements.

This method is perfectly adaptable to any inter-company co-operation (not necessarily in the setting of an extended enterprise) and to be re-used or adapted in the setting of co-operative group work. The application in an extended enterprise includes indeed all forms of inter-company co-operation; it constitutes a guarantee of a broad feasibility.

7.8. About the author

Imed Boughzala is an Associate Professor. at the National institute of Telecommunications (INT, France). His research interests include topics and issues related to Cooperative Information Systems and knowledge management.

7.9. References

[ALB 96] Alban D., *Organisation du système d'information et stratégies d'entreprise étendue*, PhD Thesis, IAE de Paris, Université Paris I Panthéon Sorbonne, June 1996

[AMI 93] AMICE, *CIMOSA: CIM Open System Architecture*, 2nd edition reviewed and augmented, Springer-Verlag, Berlin, 1993

[AUB 03] Aubertin G., Boughzala I., Ermine J.-L., "La cartographie des connaissances critiques", *Revue des sciences et technologies de l'information*, RIA-ECA series, vol. 17, no. 1-2-3/2003, pages 495–502, Edition Lavoisier

[BOU 01] Boughzala I., *Démarche méthodologique de conception de systèmes d'information coopératifs interagents pour la gestion des connaissances*, thesis, the University of Paris VI, 20 December 2001, http://www.boughzala.com/these.htm

[BOU 02] Boughzala I., "Methodology for designing interenterprise co-operative information system", *SCI'2002 : the 6th World Multi Conference on Systemics, Cybernetics and Informatics*, 14–18 July 2002, Orlando, USA

[BOU 97] Boulanger D., Dubois G., "Objets et coopération de systèmes d'information". In *Ingénierie objet: concepts et techniques*, InterEditions, May 1997

[BRO 92] Brodie M.-L., Seri C., "On intelligent and co-operative information systems: a workshop summer". *International Journal and Co-operative Systems*, June 1992

[CAU 01] Cauvet C., Rosenthal-Sabroux C., *Ingénierie des systèmes d'information*, Hermès, Paris, 2001

[CHA 00] Charlet J., Zacklad M., Kassel G., Bourigault D., "Ingénierie des connaissances: recherches et perspectives", in *Ingénierie des connaissances, évolutions récentes et nouveaux defie*, Charlet J., Zacklad M., Kassel G., Bourigault D. (eds.), Eyrolles, Paris, 2000

[DIE 00] Dieng R., Corby O., Giboin A., Goliebowska J., Matta N,. Ribière M., *Méthodes et outils pour la gestion des connaissances*, Dunod, Paris, 2000

[ERM 00] Ermine J.-L., *Les systèmes de connaissances*, 2nd edition reviewed and augmented, Hermès, Paris, 2000

[ERM 01] Ermine J.-L., *Capitaliser et partager les connaissances avec la méthode MASK, Ingénierie et capitalisation des connaissances*, Hermès, Paris, 2001

[GRU 96] Grundstein, M., "La capitalisation des connaissances de l'entreprise, une problématique de management", *Proceedings of the 5th meeting of the MCX program, Complexité: la stratégie de la reliance*, Aix-en-Provence, 4–5 July 1996

[HER 01] Hérin D., Espinasse B., Andonoff A., Hannachi C., (2001). "Des systèmes d'informations coopératifs aux agents informationnels". In *Ingénierie des systèmes d'information*, Hermès, 2001

[MAY 95] Mayer J., Benjamin P.C., Caraway B.E., Painter M.K., "A Framework and a Suite of Method for Business Process Reengineering", first part, http://www.idef.com, 1995

[MET 03] Métais E., Sèdes F., "Appariement d'informations dans les entrepôts de données: quelques approches pour le filtrage flexible", *Revue I3*, vol. 2, no. 2, Cépaduès, 2003

[OLE 98] O'Leary, D.E., "Using AI in Knowledge Management: knowledge bases and ontologies", *IEEE Intelligent Systems*, 13(3), pp. 34–39, May–June 1998

[RUM 99] Rumbaugh J., Jacobson I., Booch G., *The Unified Modeling Language Reference Manuel*, Addison-Wesley, Reading, 1999

[STE 93] Steels, L. "Corporate knowledge management". *Proceedings of the International Symposium on the Management of Industrial and Corporate Knowledge*, pp. 9–30, Compiègne, France, 27–28 October 1993

[SCH 99] Schreiber G., Akkermans H., Anjeweirden A., De Hoog R., Shadbold N., Van de Velde W., Weilinga B., *Knowledge Engineering and Management, The Common KADS Methodology*, MIT Press, Cambridge, 1999

[TAT 03] Tata S., Boughzala I., "Modelling contribution for virtual enterprise support". In *IEEE International Workshops on Enabling Technologies: Infrastructure for Collaborative Enterprise* (WETICE'2003), 9–11 June 2003, Linz, Austria

[UMM 01] *UMM, UN/CEFACT, Modelling Methodology*. Draft for trial use by UN/CEFACT WG, http://www.unece.org/cefact/, 4 April 2001

[VER 01] Vernadat F.B., "Unified Enterprise Modeling Language", *MOSIM'2001, Conférence francophone de MOdélisation et SIMulation*, Vol. 1, Troyes, France, 25–27 April 2001

[VER 96] Vernadat F.B., *Enterprise Modeling and Integration: Principles and Applications*, Chapman & Hall, London, 1996

[WFMC 99] *WfMC, Workflow Management Coalition. Terminologie et glossaire workflow*, v.2.0, http://www.wfmc.org/, February 1999

Chapter 8

Knowledge Management and Environment Scanning: A Methodological Guide to Improving Information Gathering

8.1. Introduction

In this chapter, we propose a methodological guide to improving information gathering in scientific observation processes. The method is based on the exploitation of the firm's corporate knowledge and uses an appropriate modeling to define a pertinent information gathering plan. The approach extends classical environment scanning processes, based principally on patents, towards knowledge creation and innovation processes.

Defining the requirements for information gathering is a very difficult stage in the environment scanning process. This difficulty can be explained by the fact that information needs are closely linked to our actual state of knowledge [BOU 56]. The pertinence of a definition of information needs therefore relies on our ability to explain the present state of our knowledge.

Our method leads from a partial modeling of the context and the knowledge domain to the elaboration of a requirements specification for environment scanning.

Chapter written by Thierno TOUNKARA, Institut National des Télécommunications.

The modeling of expert knowledge not only helps to determine "knowledge holes" but also helps to understand the construction experts put on their socio-economic and technological environment.

Our approach is illustrated with an application in a scanning unit at Renault[1].

8.2. Modeling of the interaction process between the corporate knowledge of the firm and its environment

In our approach [TOU 99], we consider the firm and its environment as two distinct subsystems between which there is a strong interaction.

This interaction process, which we call the "EP approach" ("Environnement/Patrimoine" in French), can be modeled into three stages (Figure 8.1 [BEN 00]):

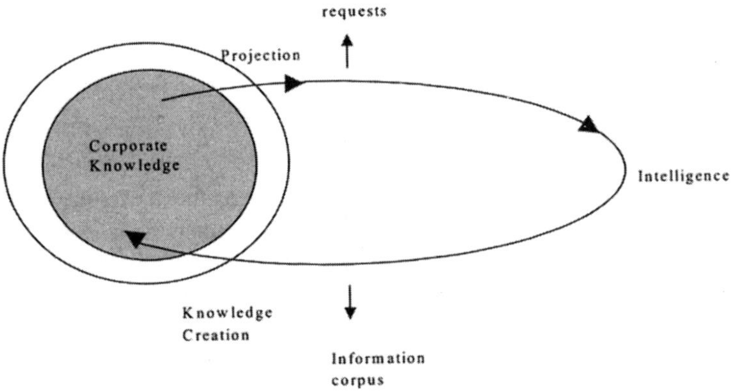

Figure 8.1. *EP interaction process*

– *Projection*: This stage involves analyzing the structure of the firm's knowledge (tacit or formalized knowledge) in relation to its environment.

– *Intelligence*: This stage moves from requests formulated in the projection stage to the elaboration of an information corpus.

– *Knowledge creation*: This is the result of the interaction process between internal knowledge and information about the environment.

In this article, we focus on the two first stages.

1 Renault is one of the leading French automobile manufacturers.

8.3. General approach

Our approach is based on the use and formalization of expert knowledge. The aim is to produce a pertinent requirements specification for the environment scanning mission in order to avoid two difficulties pointed out by Stubbart [STU 82]:

– an inability to define the information which is to be gathered;

– a tendency to reduce the scope of scanned environment, which can prevent the firm from detecting interesting opportunities.

8.3.1. *Requirements specification for environment scanning*

The requirements specification (or *information gathering plan*) is an identification of the relevant information requirements and is defined by three elements:

– *Strategic axes for the information gathering*

Here, the goal is to determine, in a appropriate way, strategic axes for the information gathering, which can be validated and classified by experts according to criticality levels. These axes "which will define the main thing that needs to be known in order to clarify the environment scanning objectives" are an expression of information gathering needs [MAR 89]. We should also note that these axes allow the structuring of the gathered corpus and thus they provide a support for making sense of the information.

– *Environment scanning "focus"*

An environment scanning "focus" relates to further information that has to be gathered for each strategic axis or combination of axes in order that the environment scanning goals can be attained. The foci are weighted by experts and classified according to their importance on each axis. Then, using the foci and information sources, it will be possible to develop relevant requests for documentary information gathering.

– *Information sources*

These make up an inventory of formal and informal sources relating to the strategic axes. Information sources are classified according to their type and their reliability level. This inventory will contribute to the structuring of a network of internal actors for information analysis.

8.3.2. *Description of the approach*

Figure 8.2 is a global description of the projection stage in our approach.

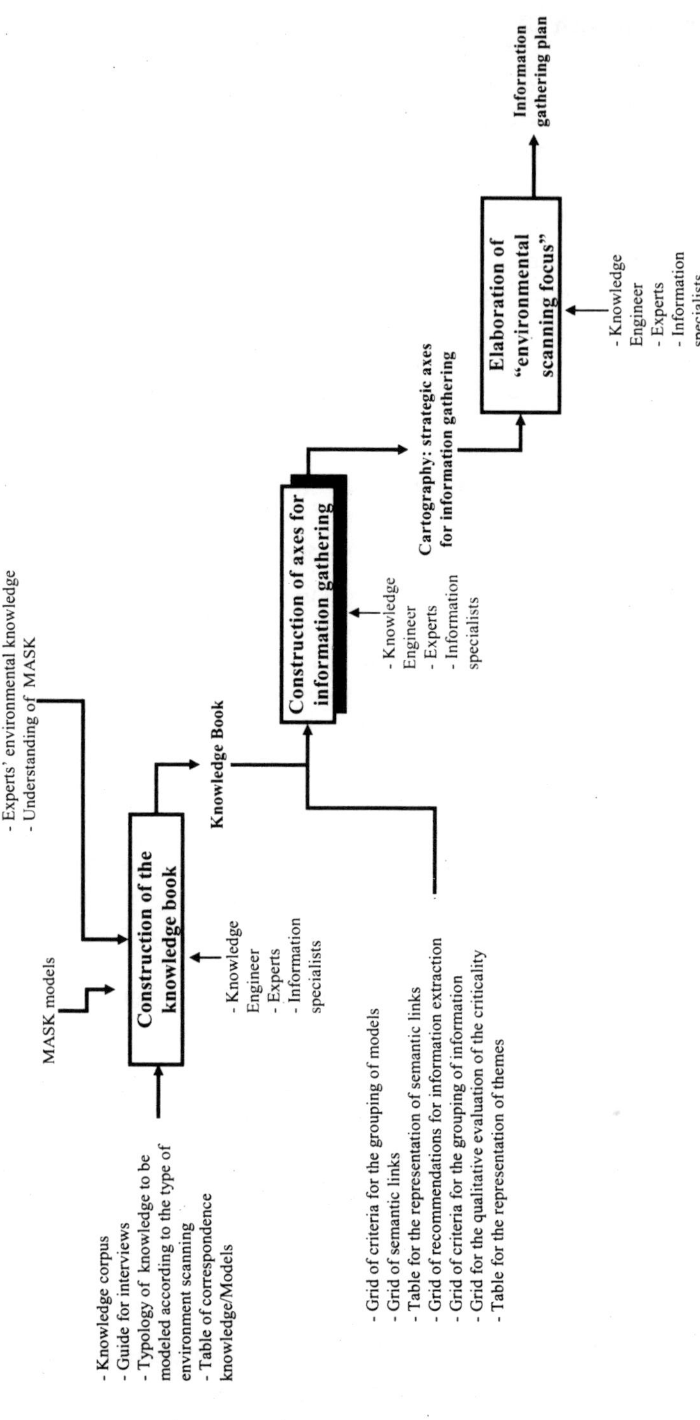

Figure 8.2. *Activity model for the projection*

The methodological guide we propose can be described in terms of three actions:

– *Construction of a knowledge book*

Using knowledge engineering methods [CHA 00], we model part of experts' knowledge so as to clarify the environment scanning goals.

– *Construction of strategic axes for information gathering*

From the knowledge book elaborated on the basis of expert knowledge and using an abstraction mechanism, we determine strategic axes that are an appropriate expression of the information gathering requirements.

– *Elaboration of the environment scanning "focus"*

Using the structuring of the environment into strategic axes, we list, validate and weight "environment scanning foci".

8.4. Knowledge book

Firms are becoming more and more involved in knowledge capitalization actions in order to overcome a loss of know how and to optimize their productivity [DRU 92] [GRU 99].

These different actions are possible thanks to a process that transforms tacit knowledge into codified knowledge. This process, called "Externalization", is important for organizational knowledge creation [NON 97].

The modeling work must be done by knowledge engineers, who collect and formalize in an exploitable way parts of the experts' knowledge. The result of this formalization is a group of models that we call the knowledge book. In our approach, this knowledge book is used for structuring the environment in terms of strategic axes for information gathering.

If we want to have a relevant structure, we must avoid relying on distorted perceptions of the environment and therefore we must construct these perceptions in an appropriate way. That's why Weick [WEI 95] considers that environment is built by individuals. This construction matches a concept developed by Waterman [WAT 90], which he called "*structuring of the unknown*".

The modeling of experts' knowledge not only helps to identify "*knowledge holes*" (or "*knowledge deficiencies*"), but also helps to understand the *constructions that experts use to describe their socio-economic and technological environment.* All these elements are used to define the needs for environment scanning.

Experts, thanks to their "environmental knowledge", will be able to help us build an appropriate representation of their environment. This knowledge is made up of "factual knowledge" (strategic movements of competitors, evolution of critical technologies, domain maturity) as well as knowledge about information sources and the rules that govern the functioning of the environment [BEL 00].

Factual knowledge is characterized by the stimuli that individuals identify in their environment [STAR 88] or "interruptions" [WEI 95] that they evaluate according to various characteristics. These stimuli are then integrated into general interpretation processes.

The Knowledge Book we build thus takes into account experts' knowledge of the environment.

8.4.1. *Methodological tools*

Methodological tools make construction of the knowledge book easier. They are listed in Table 8.1.

Methodological tools for MASK application	**Methodological tools concerning the different types of knowledge to be modeled**
– MASK models – Guide for interviews	– Typology of knowledge to be modeled according to the type of environment scanning – Table of correspondence between knowledge to be modeled and MASK models

Table 8.1. *Methodological tools used for construction of the knowledge book*

8.4.1.1. *Methodological tools for MASK application*

8.4.1.1.1. MASK models

The MASK method (Method for Analysis and Structuring of Knowledge) [ERM 00] was applied for the first time at CEA[2] in 1993 [LEB 94]. With MASK, knowledge is collected by consulting "knowledge sources" within the organization (experts, specialists or documents).

2 The French Atomic Energy Commission (CEA) is a key player in research, development and innovation in the fields of energy, defence, information technologies, communication and health.

A good understanding of MASK models is necessary; these models are a guide for the knowledge engineer during interviews with experts and they make easier the formalization of the collected knowledge.

MASK uses several models, which correspond to various ways in which a domain can be viewed:

– *the domain model*: a description of general phenomena;

– *the activity model*: breaking down an activity into stages; these stages are linked by exchanges of data, material flows, etc.;

– *the concepts model*: a classification of concepts and objects in the domain;

– *the tasks model*: a representation of the strategy (know how) used to solve identified problems concerning the knowledge system studied;

– *historical and line models*: descriptions of the evolution of a knowledge group; the goal is to have a better understanding and mastery of the evolution of the knowledge laws.

8.4.1.1.2. The guide for interviews

This guide gives useful recommendations concerning:

– difficulties the knowledge engineer may have during interviews (these difficulties can relate to recorded interviews, to the choice of place, to communication with the expert, etc.);

– the way to begin the first interview with the expert because, for the expert, this will be his or her first contact with a capitalization method and a poor understanding of the method can lead to major problems;

– the choice of the relevant type of interview according to the context and the preferences of the knowledge engineer (free, directed or mixed interview);

– the way to correct and validate models that are elaborated after each interview with the expert.

8.4.1.2. *Classification of the knowledge to be modeled according to the type of environment scanning*

We can classify environment scanning into three types that depend on the goals of the scanning action, the modes of action and the horizon targeted [DEG 01], [BOU 90]:

– "Spontaneous" environment scanning is concerned with a precise and concrete subject; it is spontaneous because the scanning provides, in a natural way, answers relating to a current issue and corresponds to an identified and formulated need.

– "Reactive" environment scanning allows, using an organized approach, the understanding of evolution in a specific area and the detection of opportunities and threads.

– "Long-term" environment scanning: the goal here is to have a vision of the future over a probable horizon/time and also to remove uncertainty about the direction of evolution.

Using this classification and feedbacks from scientific observation projects realized at CEA [TOU 99] and at Renault [TOU 01], we propose a typology based on the type of environment scanning and the type of knowledge to be modeled in the knowledge book (Table 8.2).

	Knowledge related to the firm's activities	**Environmental knowledge**
"Spontaneous" and "reactive" environment scanning	Knowledge about the functioning of the organization Knowledge about internal businesses/activities Knowledge about technologies used by the firm Knowledge about the limits of technologies used in the firm Knowledge about technological needs evolution in the firm Knowledge of the business terminology Knowledge about the physico-chemical principles used in technologies	Knowledge about the evolution of the domain Knowledge about technologies used or planned in the external environment (by competitors, suppliers, etc.) Knowledge about the direction of evolution of technologies that is critical for the firm's activities (trends, evolution laws, etc.) Knowledge about strategic movements by competitors Knowledge about information sources
"Long-term" environment scanning	Knowledge about internal businesses/activities Knowledge about technologies used by the firm Knowledge of the business terminology Knowledge of the history evolution of the domain	Knowledge of the history of the evolution of the domain Knowledge about domain maturity Knowledge about information sources

Table 8.2. *Typology of knowledge to be modeled according to the type of environment scanning used*

We show in case studies [TOU 02] that MASK models are appropriate for the modeling of the different types of knowledge listed in Table 8.2. We therefore propose a table of correspondence between knowledge to be modeled and MASK models.

8.4.2. *Case study: Renault*

The goal of the "automobile pollution control" scanning unit at Renault is twofold:

– broadening knowledge about the "treatment of nitrogen oxide (NO_x) produced by diesel engines". Nitrogen oxides are gaseous pollutants and controls concerning their emission are very strict;

– evaluating the approach we have presented in this chapter by a comparison with classical processes of environment scanning.

We interviewed one expert so that we could construct the knowledge book: there were five interviews of two hours each. We used two types of model: domain models and concept models. Table 8.3 shows a detailed plan of the phenomena and concepts modeled. For confidentiality reasons, the only parts of the results of the study are presented.

Domain models (MD)	Concepts models (MC)
• **MD_1** – Direct reduction of NO_x • **MD_2** – Harmful effects due to the reaction between hydrogen and oxygen • **MD_3** – Adsorption of NO_x by the active site of the catalytic converter • **MD_4** – Harmful effect due to the adsorption of sulfates by the active site of the catalytic converter • **MD_5** – Reduction of sulfates • **MD_6** – Electrochemical reduction of NO_x	• **MC_1** – Pollution control systems

Table 8.3. *Extract from the knowledge book structure*

8.5. Construction of axes

The aim is to extract from the knowledge book the main directions for appropriate information gathering. This process, as a result of cognitive operations of individuals (experts and knowledge engineers) on the models, transfers to an upper layer some aspects so that new combinations can be obtained and strategic axes for information gathering deduced based on these: Piaget calls this process "abstraction" [PIA 77]. The process is shown in Figure 8.3. A similar process is found in the "puzzle" method of Lesca [LES 95]; it is a method for the analysis of weak signals.

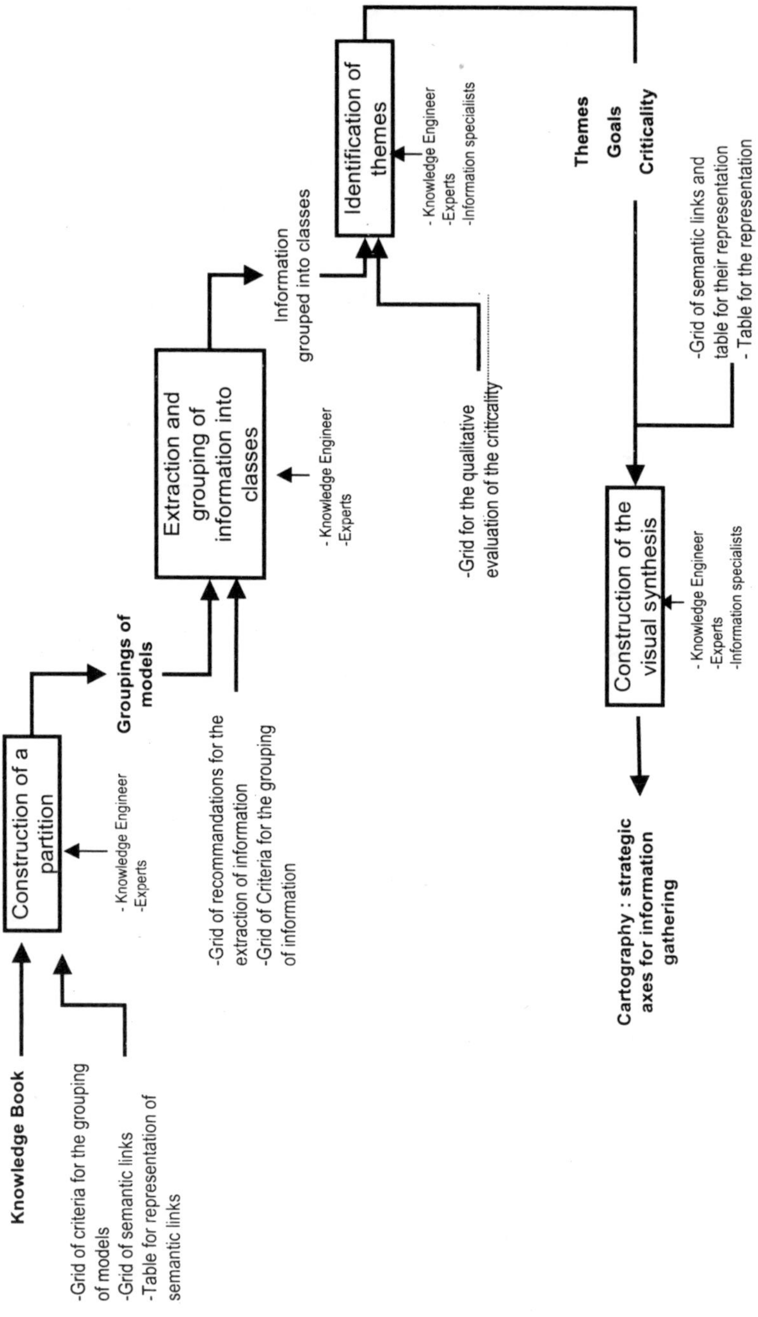

Figure 8.3. *Recommendations for the construction of strategic axes for information gathering*

8.5.1. *Construction of a partition*

The construction of a partition consists of grouping the models of the knowledge book into significant classes. The aim is to make easier analysis of the knowledge book for the development of the information gathering axes.

Furthermore, the complexity of the analysis is reduced when models are grouped into categories.

The dictionary *Le Petit Larousse* [LAR 02] defines partition of models of the knowledge book as "mutually exclusive classes of models; the union of these classes must match the complete models of the knowledge book".

8.5.1.1. *Methodological tools for the elaboration of a partition*

These methodological tools are essentially grids that actors can use to construct and validate the classes which compose the partition.

8.5.1.2. *The grid of criteria for the grouping of the models into classes*

The choice and the construction of these criteria are based on feedback from the application of our approach in real cases, as well as on theoretical foundations concerning:

– the grouping of information for the meaning of construction;

– the concepts of classes, generalization and generic class (or super class) used in computer science [MUL 00];

– the elaboration of scenarios for problem solving; Altshuller's studies [ALT 99] in that domain provide a reference.

We propose three types of grouping based on the following criteria: analogy, proximity and "connectivity" (Table 8.4).

In the grid, we also propose methodological elements (generic model, semantic link) which support the validation of the different types of grouping: these are the "*elements of validation*".

Grouping criteria	Definitions	Elements of validation
Analogy: grouping guided by a generic model	Models are grouped on the basis of underlying similarities which can be analyzed in a more general model that we call the "*Generic Model*". The generic model is an abstraction of all models making up the grouping. Models can thus be grouped: – on the basis of known scientific phenomena (chemistry, biology, physics, etc.): there is a direct analogy; – on the basis of images, symbols, legends, myths (an indirect analogy).	***The generic model*** We suggest, for each grouping based on analogy, to elaborate a generic model that will be used as a: – support for the validation of the grouping; – guide for the extraction of information. Generalization is a difficult operation and it needs a good abstraction ability.
Proximity: grouping guided by common information	Proximity is a criterion that is less restrictive but rougher than the analogy criterion. Here, we need only to find a common characteristic (or information) to group models. This criterion is very subjective: the risk can be reduced if the grouping process is led collectively by a group.	
Connectivity: grouping by construction of a scenario based on semantic links between models	With the connectivity criterion, the grouping is a set of models that can be connected by a chain of semantic links (causal, influence and contradiction links). This chain of semantic links must describe a scenario aimed at resolution of the problem which has led to the scanning action. This type of grouping is appropriate in the case of "*reactive environment scanning*" (see section 8.4.1.2).	***Significant links*** We suggest a graphical representation of the existing links between models of the grouping (Table 8.5). These links will be used: – to validate the grouping; – to make the extraction of information easier. The different types of semantic links that we can use to connect models are described in Tables 12 and 13.

Table 8.4. *Grid of criteria for the grouping of models*

8.5.1.3. *The grid of semantic links*

This grid is essentially a support for the construction and validation of the groupings by "connectivity". It can also be used for visual synthesis construction (see section 8.5.4).

Referring to the studies of Caron [CAR 97] and to TRIZ methodology [ALT 96], we propose three types of semantic links (Table 8.5) for making connections between models.

Type of link	Definition
Causal link	Information X will be linked to information Y if the cause of Y is X. Causal links are very strong and, as noted by Lesca, represent an ideal that is almost inaccessible.
Influence link	Information X will be linked to information Y if X influences Y without being the unique and/or direct reason of the existence of Y.
Contradiction link	In cases where we cannot easily link information by causal or influence links, we can try to find a *contradiction* between them.

Table 8.5. *Grid of semantic links*

Causal and influence links can be characterized by "positive", "harmful" or "unevaluated" actions. The aim of these qualifications is to make sure that the grouping by connectivity is really guided by the resolution of the problem that has leaded to the scanning action (see section 8.5.1.2).

We assign a precise graphical representation to each type of semantic link (Table 8.6).

Links	Graphical representation
Causal link unevaluated	
Positive causal link	
Harmful causal link	
Influence link unevaluated	
Positive influence link	
Harmful influence link	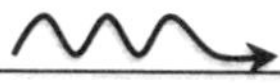
Contradiction link	

Table 8.6. *Table for the representation of semantic links*

8.5.1.4. *Renault case study: an example of grouping by connectivity*

In Renault case, we elaborate two groupings using the *connectivity* criteria (Figure 8.4):

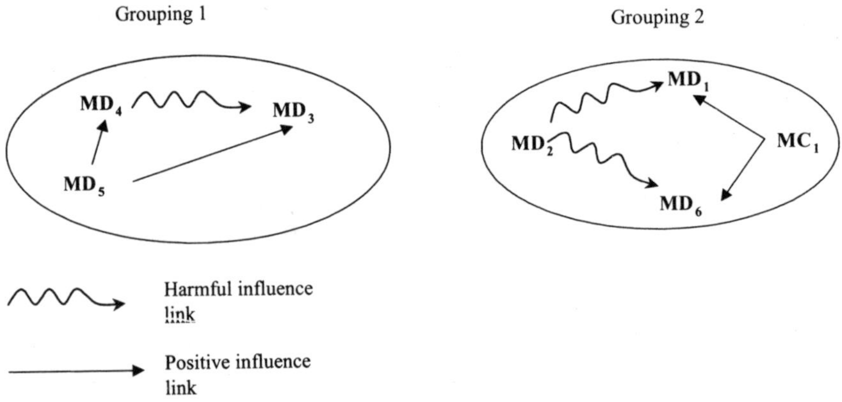

Figure 8.4. *Partition obtained in the Renault case study; MD_i and MC_i refer to models in Table 8.3*

Description of the scenario explaining the first grouping

One of the listed problems is "the harmful effect due to the adsorption of sulfates by the active site of the catalytic converter". In a diesel engine, sulfur in the fuel is adsorbed by the catalyst in the form of sulfates. The catalyst sites reserved to nitrogen oxides (NO_x) are then polluted. The consequence is a decrease in *the adsorption efficiency of NO_x by the catalyst* (*harmful consequence*).

Hydrogen (H_2) injection, in the catalyst, is a possible solution that eliminates the sulfates by a chemical reaction (sulfate reduction) (*positive action*) and therefore can solve the problem of the poisoning of the NO_x sites by sulfates (*positive consequence*).

8.5.2. *Extraction and grouping of information into classes*

The goal is to extract, from each grouping of models, information that we must complete by a deeper search to fill the "knowledge holes". This information is synthesized and analyzed to constitute strategic axes (or themes).

In the puzzle method [LES 95], the items of information to be completed are grouped into classes. We use similarity, proximity or analogy criteria to elaborate the classes.

In the Renault Project, we obtained three classes of information; they are presented in Table 8.7.

Class 1	Class 2	Class 3
– NO_x reduction by hydrogen – Electrochemical reduction of NO_x in N_2 or NH_3 in an electrolytic environment (H^+ or O^{2-}) – NO_x reduction by ammonia – Bringing nearer hydrogen (H_2/H^+) and NO_x in a catalytic converter – etc.	– Desorption and reduction of sulfates in a NO_x trap – Desorption and reduction of sulfates in a NO_x trap by injection of H_2 or by production of H^+ – NO_x trap without affinity for sulfates – etc.	– Technology of pollution control allowing NO_x reduction by hydrogen (H_2) – Technology of pollution control using a NO_x trap with an injection system of H_2 – Technology of pollution control using a NO_x trap with an electrocatalytic cell – Technology of pollution control allowing NO_x reduction by an electrochemical method – Technology of pollution control using a $DeNO_x$ catalyst with an H_2 injection system

Table 8.7. *Summary of classes of information obtained*

8.5.3. *Identification of themes*

The aim is to constitute, from classes of information, appropriate themes for information gathering. The criticality of these themes must also be qualitatively evaluated: this is the criticality analysis.

8.5.3.1. *Qualitative evaluation grid for the criticality of themes*

We elaborate this grid using two criteria:

– newness of the theme in comparison to the themes considered by the expert or the expert's team before the construction of the knowledge book;

– pertinence of the theme according to the environment scanning goals.

8.5.3.2. *Approach and illustration*

The stages are as follows:

– giving a significant name to each class; the name must be representative of information composing the class, i.e. its theme;

– defining for each theme the information gathering goals;

– evaluating the criticality of each theme with experts: the qualitative evaluation grid can be used (Table 8.8).

Newness of the theme	Pertinence of theme
– new – already present	– essential – interesting – not pertinent – unevaluated

Table 8.8. *Qualitative evaluation grid for criticality*

This is an example from Renault case:

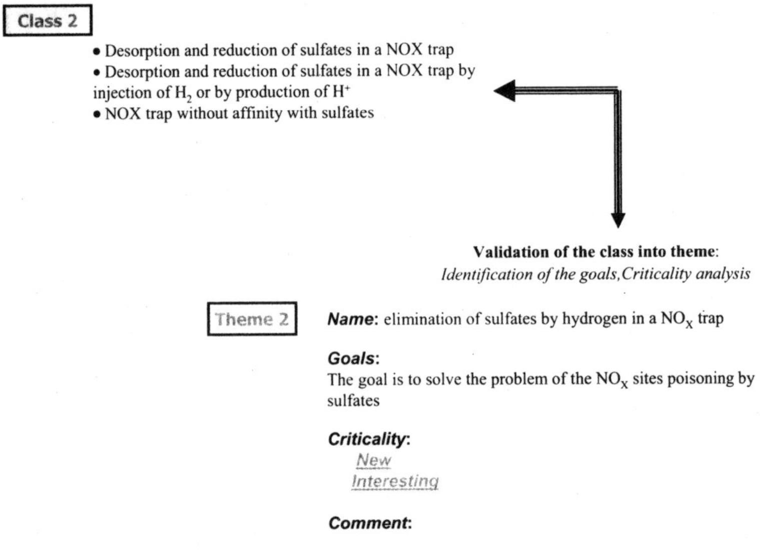

Figure 8.5. *Example of the validation of class into theme*

8.5.4. *Construction of the visual synthesis*

The goal of the visual synthesis construction is to replace many "literal" items of information by *synthesized and collectively intelligible information.*

In the Renault case, classes 2 and 3 of Table 8.7 were validated into themes for the information gathering (they became respectively themes 2 and 4 in Figure 8.6). The criticality analysis with experts led to the splitting of class 1 into two classes in order to highlight a new and essential theme: theme 3.

Thus, we obtained four strategic axes for information gathering, which are shown in Figure 8.6.

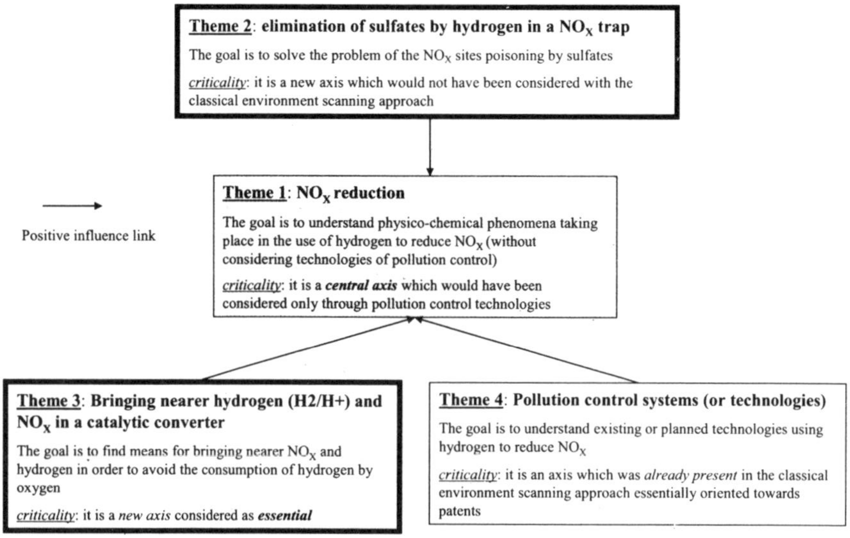

Figure 8.6. *Cartography of strategic axes for information gathering*

8.6. Elaboration of the "environment scanning focus"

The goal is to select, on each strategic axis, information that has to be supplemented by further information in order to highlight the environment scanning goals: we call this further information the "environment scanning focus".

The additional information must be worded so as to make formulation of requests easier.

The elaboration of the "environment scanning focus" needs the participation of information specialists who are responsible for searching the information sources.

Some of the results in the Renault case concerning the environment scanning focus are presented in Table 8.9.

Axes	Environment scanning focus
Theme 1	– … – NO_x reduction by ammonia: temperature, gas pressure and reaction stability conditions; ammonia/ NO_x ratio
Theme 2	– Desorption and reduction of sulfates in a NO_x trap – Desorption and reduction of sulfates in a NO_x trap by injection of H_2 or by production of H+ – NO_x trap without affinity for sulfates – …
Theme 3	– … – Pollution control technologies allowing gas deviation: *bypass* system, etc. (tight shut-off valve in particular); *This point was added by experts*
Theme 4	– … – Pollution control technologies allowing NO_x reduction by an electrochemical method – … – Pollution control technology using a $DeNO_x$ catalyst with an electrocatalytic cell

Table 8.9. *Examples of focus in the Renault case*

After the development of the foci, we can define formal and informal information sources.

We indicate on each strategic axe of the cartography:

– formal information sources (databases, etc.);

– internal actors who have interesting information and/or who are able to treat information that has been gathered;

– competitors, suppliers, research centers, etc.

This task must be led by both experts and information specialists.

8.7. Evaluation of our approach

The project carried out with the "automobile pollution control" scanning unit at Renault has allowed us to make a first evaluation of our approach.

The information gathering was organized in collaboration with the Documentary Information Centre of Renault. We obtained a corpus of 200 documents (abstracts) from 11 technical and scientific databases.

After a first sorting, 10 non-relevant documents were eliminated. The sorted corpus was given to experts, who confirmed the relevance rate of the documents gathered.

8.8. Conclusion

The goal of the work we present in this chapter is to optimize the efficiency of environment scanning processes in firms in order to support innovation processes.

We have tried to highlight the added value of the use of corporate knowledge modeling as a support for information gathering. We show, in the Renault case, that a partial modeling of experts' knowledge brings more efficiency in the projection stage, gives a better structuring of the environment scanning domain and thus a more relevant requirements specification for information gathering.

The methodological guide we propose complements existing prospective methods, which are based on the exploitation of experts' knowledge for defining information needs. Our method allows the use of knowledge engineering techniques to do the following:

– structure the exchange between the expert and the knowledge engineer when defining information needs;

– restore the high level of expertise after interviews with the expert;

– reformulate the environment scanning problem in order to tackle it in new ways.

Next, we intend to design a collaborative system that will combine the different methods and tools used in the guide (MASK models, tables, grids, cartography tools, etc.). The aim is to make the methodology easier to apply in firms. We envisage this collaborative system as a working space where the different actors (experts, information specialists, knowledge engineers) will interact in order to elaborate and validate the requirements specification for the environment scanning.

8.9. References

[ALT 96] ALTSHULLER G., *And Suddenly the Inventor Appeared*, Technical Innovation Center, Worcester, MA, 1996.

[ALT 99] ALTSULLER G., *The Innovation Algorithm, TRIZ, Systematic Innovation and Technical Creativity*, Technical Innovation Center, Worcester, MA, 1998.

[BEL 00] BELMONDO C., *L'articulation entre outils de gestion et connaissances tacites/explicites et leur intégration dans la gestion des connaissances: application à une cellule de veille concurrentielle*, AIMS, May 2000.

[BEN 00] BENHAMOU P., ERMINE J.-L., SABROUX C., ROUSSEAU F., TOUNKARA T., *Une méthode intégrant les activités de gestion des connaissances et de veille*, Proceedings of the IC2000 Congress, Ingénierie des Connaissances 2000, Toulouse, France, 10–12 May 2000, pp. 117–130.

[BEN 01] BENHAMOU P., ERMINE J.-L., TARAN J.-P., TOUNKARA T., WAETERS A., *Evolution des connaissances et innovation: application à une technologie laser à l'ONERA*, EGC2001, Nantes, 18-19 January 2001.

[BOU 56] BOULDING K.E., *The Image*, Arbour Paperback, University of Michigan, 1956.

[BOU 90] BOURCIER-DESJARDINS R., MAYERE A., MUET F., SALAUN J.M., *Veille technologique: revue de la littérature et étude de terrain*, CERSI, 1990.

[CAR 97] CARON-FASAN M.-L., *Veille stratégique: création de sens à partir de signaux faibles*, PhD thesis in Science Management, University of Grenoble II, 1997.

[CHAR 00] CHARLET J., ZACKLAD M., KASSEL G., *Ingénierie des connaissances, évolutions récentes et nouveaux défis*, Eyrolles, 2000.

[DEG 01] DEGOUL P., *Essai sur l'articulation entre intelligence économique et gestion des connaissances*, EGC2001, Nantes 18-19 January 2001.

[DRU 92] DRUCKER P., *The new society of organizations*, Harvard Business Review, pp. 95–104, September-October 1992.

[GRU 99] GRUNDSTEIN M., ROSENTHAL-SABROUX C., *Système d'information et capitalisation des connaissances: un enjeu nouveau pour la conception de système d'information*, Prédit, October 1998.

[ERM 93] ERMINE J.-L., "Génie logiciel et génie cognitif pour les systèmes à base de connaissances", *Collection Tec et Doc*, Lavoisier, Paris, 1993.

[ERM 00] ERMINE J.-L., *Les systèmes de connaissances*, Editions Hermès, Paris, 1996, 2nd ed, 2000.

[LEB 94] LEBLANC H., DAGORN D., ERMINE J.-L., *Knowledge Management in Human Radioprotection*, ISMICK'94, Compiègne, 26-27 October 1994.

[LES 95] LESCA H., CARON M.-L., "Veille stratégique: créer une intelligence collective au sein de l'entreprise", *Revue Française de Gestion*, September-October 1995.

[MAR 89] MARTINET B., RIBAULT J.M., *La veille technologique, concurrentielle et commerciale*, Edition d'Organisation, 1988.

[MUL 00] MULLER P.-A., GAERTNER N., *Modélisation objet avec UML*, Eyrolles, 2000.

[NON 97] NONAKA I., TAKEUCHI H., *La connaissance créatrice: la dynamique de l'entreprise apprenante*, De Boeck Université, 1997.

[PIA 77] PIAGET J., *Recherches sur l'abstraction réfléchissante*, Presses Universitaires de France, 1977.

[ROU 99] ROUSSEAU F., ERMINE J.-L., *Processus d'interaction entre le patrimoine scientifique et technique d'une organisation et son environnement*, SCIP France, June 1998.

[STAR 88] STARBUCK W.H., MILLIKEN F.J., "Executives perceptual filters: what they notice and how they make sense", in Hambrick D. (ed), *The Executive Effect: Concepts and Methods for Studying Top Managers*, JAI Press, 1988.

[STU 82] STUBBART C., Are environmental scanning units effective, *Long Range Planning*, vol. 15, no. 3, pp. 139-145, 1982.

[TOU 99] TOUNKARA T., *Gestion des connaissances et veille*, *rapport de DEA*, LAMSADE, Paris-Dauphine University, September 1998.

[TOU 01] TOUNKARA T., BENHAMOU P., DEVALLAN F., ERMINE J.-L., ROSENTHAL-SABROUX C., *Gestion des connaissances et Veille: vers un guide méthodologique pour améliorer la collecte d'informations*, Actes du congrès VSST2001, Barcelona, 15-19 October 2001.

[TOU 02] TOUNKARA T., *Gestion des connaissances et Veille: vers un guide méthodologique pour améliorer la collecte d'informations*, PhD Thesis, Paris-Dauphine University, December 2002.

[WAT 90] WATERMAN R.H., *Adhocraty: The Power to Change*, Whittle Direct Books, 1990.

[WEI 95] WEICK K.E., *Sensemaking in Organizations*, Sage, 1995.

[WIE 67] WIENER A., KAHN H., *The Year 2000*, MacMillan, New York, 1967.

Chapter 9

The Concept of "*Ba*" within the Japanese Way of Knowledge Creation

The concept of *ba* was introduced in 1996 by Ikujiro Nonaka and Noburo Konno. Since then it has played a major role in the Japanese way of knowledge creation. It now belongs to the specialized jargon of knowledge management that has emerged from the Japanese archipelago, which is different from the IT-oriented approach that has emerged from the United States.

The characteristic of Japanese culture in this concept makes it hard to find in Western languages a unique word that is equivalent, clear, distinct and without any other implications. Therefore this chapter proposes using the formulation *Strategic Knowledge Community* as an equivalent.

This contribution is organized in three parts. First of all, it strives to define the *ba* concept from a Japanese cultural point of view. Then, it considers some philosophical implications of the concept. Finally, it presents some case studies from the Human Health Care program in the Eisai Company.

This chapter was made possible because of an investigation program about the Japanese way of Knowledge Creation supported by the Science and Technology Service of the French Embassy in Japan. The overall results of this investigation will be published in 2006 [FAY 06].

Chapter written by Pierre FAYARD, Director of Cendotec (Franco-Brazilian Centre for S&T Documentation), Sao Paulo, Brazil, and Professor at the University of Poitiers, France.

9.1. A Japanese concept

Ba is a *Kanji* ideogram, the left part of which means ground, boiling water or what is rising and the right part of which means to enable. On the one hand, it denotes a potential and, on the other, a kind of engine that gives a direction. One may talk about a *good ba* when relational situations energize people, making them creative with positive and dynamic interactions. The right part of the ideogram refers to the *yin* and *yang* philosophy of permanent transformation [JAV 02].

As Kitaro Nishida wrote: "Reality is a succession of events that flow without stopping" [NIS 90]. The use of the ba concept comes from this philosopher, who identified a physical space in which a hidden power is lying, from which one can receive energy when one dives in. However, it is not just a place but also a moment in which one may undergo a transformation and emergent process. A *ba* can be memorized and opened up to a continuity of relationships within a kind of atmosphere that refers to a particular feeling linked to a community's shared space and time. "What we call time, space and material force are simply concepts established in order to organize these facts and to explain them" [NIS 90]. From this perspective, a *ba* can be identified with a kind of level of consciousness.

For Ikujiro Nonaka, a *ba* "could be thought as a shared space for emerging relationships. This space can be physical (e.g. office, dispersed business space), mental (e.g. shared experiences, ideas, ideals) or any combination of them. What differentiates *ba* from any ordinary human interactions is the concept of knowledge creation. *Ba* provides a platform for transformation and integration of knowledge. *Ba* may also be thought as the recognition of the self in all. According to the theory of existentialism, *ba* is a context which harbors meaning. Thus, we consider *ba* to be shared space that serves as a foundation for knowledge creation" [NON 98].

As a positive and voluntary field of constraints, *ba* is favorable to constructive human interactions between individuals and between them and their usual environments. Exchanges of data, information and opinion, collaboration and mobilization on a project to meet both requirements and the unknown convey the idea of *ba* within an organization. It can be understood as emptiness appropriated for emergence or as a kind of "oriented but not specifically determined" open, tacit and consensual space. *Ba* does not come into existence as a result of regulations and it differs from the *command and control* model of traditional pyramidal management. On the contrary, it comes into existence as a result of voluntary membership within an *energize and stimulate* mode through care and mutual respect. *Ba* is fundamentally subjective and relational and one becomes involved in it because it is ruled by common interest and because there are no conflicts within human relationships.

As far as the four stages of the SECI (Socialization – Externalization – Combination – Internalization) model of Nonaka are concerned, it is possible to consider *ba* through different particularities: its emergence, the socialization it provides, the systematic interaction it allows and finally its effect as an agent for internalization. *Ba* includes a tacit component when emotions, experiences, feelings and mental images are shared. It provides a context for socialization and an existential space in which individuals can transcend the limits imposed by the physical implications of the broad spectrum of their capacities. Nonaka used to say that care, love, confidence and responsibility are required. In addition to this inter-individual relationship, a collective relationship allows practices, values, processes, culture and climate to be shared in a more or less formalized way.

Virtual *ba* may function using information and communication technologies and remote networking. Interactions combine the tacit and the explicit within knowledge spirals. Finally, *ba* provides an adequate context for internalization of knowledge and catalyses reflection so that it is actually transformed into action! Effective *ba* can be revealed in various ways. For example, while one is passing through a store, weak signals could be perceived and then combined and enriched with other data, hypotheses about markets, purchase attitudes or scenarios about the future …

Ba is what allows scrupulous observers to be tuned into the right rhythms in order to take efficient decisions in terms, for example, of supplies, answers to questions or ways to present services and products. Interaction with consumers and users may also create global *ba*. Relationships within a *ba* do not exist *a priori*; they are not predetermined or come from any external physical model of human implications. The inner coherency of *ba* reveals itself through organic interactions based on vision and community knowledge effort rather than as the effect of a mechanical concentration produced by a dominant centre. These interactions lead to an idea of a *higher self* [NON 98] and continuous exchanges strengthen inner relationships. Individuals create the *ba* of teams, which in turn create the *ba* of organizations.

9.2. "Elementary, my dear Watson!"

To illustrate the complementary difference between, on one hand, the perception of weak signals produced by serendipity, fuzzy receptivity, tacit know-how, craft experience and intuition, and, on the other hand, explicit objective information, Noburo Konno makes reference to the pairing *Holmes–Watson*. Sensible, clear-sighted and creative, Sherlock Holmes identifies weak signals and then processes them by induction and combines them to make sense. In contrast, the deductive Dr Watson analyses and that is the reason why he finds out later. He is rationalizing like a historian what Holmes presents to him as "elementary!".

These two ways to process are not antagonist but complementary. Dr Watson's preconceptions orient Holmes's perceptions and even induce the sensibility that leads him to identification of relevant signals. In doing so, Dr Watson's rationality frees the imagination of Holmes, making him capable of devoting all his time and art to his insight activity. It constitutes the backdrop, the previously accessible and explicit knowledge and competences that complement Holmes' creativity and availability. Through each investigation, the Holmes–Watson pair creates a *strategic knowledge community* set up with the aim of discovering the truth! They bring together partners and all possible indicators that allow them to move ahead to achieve their aim and thus meet their detective commitments.

To highlight the philosophical dimension of *ba*, Noburo Konno quotes Derzou Ouzala, the central character in a Kurosawa film. This lonely trapper from deep inside Siberia develops a high sensitivity to signals that are coming from his environment and the harmony in which he is living with his conditions and circumstances makes him part of nature itself. "Subject and object do not exist separately for they are the two sides of one reality" [NIS 90]. Derzou fully communicates with this whole, in which he is just a moment, a modality within which he is not isolated as an autonomous and self-sufficient being. If the time comes for him to die, he dies and stays in harmony with what the environment requires. On the other hand, as long as he retains the least of his energy, he is still active without distinguishing himself from what is surrounding him and in which he is participating.

Derzou Ouzala does not constrain events but merges himself within the global unity and finds the best solutions that result from the pairing of action (himself) with reaction (the environment). Walking, he experiences the ground's reaction and does not sink because the ground is part of the movement within a complementary interdependent unity. This environmental and relational comportment is based on community. Derzou Ouzala is never alone and isolated from the world. The same energy underlies nature and makes him alive within an included *ba* at the very junction between the past and the immediate future. It is not he who wants and desires but the very community that he constitutes with nature.

In a similar way, Im Kwon-taek's film *Wild with Women and Painting*[1] relates episodes from the life of the famous Korean artist Ohwon – Jang Seung-Up – and stresses the intimate relationship that unifies beings, irrespective of forms and particular existences. Throughout successive scenes, Ohwon absorbs the energy that comes through the river's flows, the hustle and bustle of foliage, a swarm of hundreds of birds, the deaf snow that covers a landscape … till he ends up by disappearing inside the inferno of a ceramist's kiln! There is little difference

1 Original title: *Chiwaseon*, South Korea, 2001.

between his art and the work of nature. The achievement of his painting translates the permanent flow of transformations. In other words, because the artist resounds with them, he becomes able to leave them revealed as an echo through his paintings. Women and wine represent the means to dissolve the limits of his self-increasing porosity so that he maintains a qualitative sensibility that opens him to an intimate communion with this higher reality. His creativity relies on his ability to be one self with the whole and to vibrate with the forces that underlie it. Through his very exposure to nature's forces, Ohwon strives to awake similar modalities within himself so that he can translate them onto paper. That is the essence of his work and the artist does not act just relying on his will alone.

Let us talk now about the firm. Because it remains in tune with the market and avoids barriers that slow communication and adaptation with the market, the firm does not consider itself as exterior. It takes profit from the energy of the evolutions of the market as long as it satisfies the market's needs. An inclusive *ba* makes it possible to maintain flexible availability that orients itself according to the perceived signals, which are converted into information and then into knowledge within the community atmosphere. The main danger is to be cut off from the reality that flows. When something new is created, the author of it will contribute energetically to this new existence and may lose focus on everything else. What is all around may be no longer perceived.

"To get an idea means making a choice, having a bias, and so leaving in the shadow a part of reality: developing a partial thinking and so a biased one [...] the wise man has neither rules nor principles, he is ready to choose one issue or the other depending on the moment. He has an opportunistic capability advisedly and does what the situation requires" [JUL 02]. As for failure, says Noburo Konno, thinking about it may just create conditions that invite it. "The environment is infernal because hell lies inside you though you become part of it"[2]. The cycle of transformations is endless. The hand inside the flow cannot stop it, as with time. Flow is the keyword and so it is for information. "If you keep it enclosed, you poison yourself as a cancer does. You may be killed by information if you keep it!"[3] *Ba* philosophy has nothing to do with an out-of-context assumed creation of knowledge in an individual way, autonomous and excluding any human interactions. In contrast, it opens up dynamic process that surpass individual limits and it comes to reality through a *platform* where a common language is used to achieve community aims and goals. *Ba* is focused on the *knowledge front* and the human energy it uses can be extended and optimized with the capabilities of ICT (Information and Communication Technologies). ICT cannot be the first step in KM. Flowing without frontiers, *ba* modifies itself depending those who participate.

2 Noburo Konno, interview with the author of this chapter, 2002, Tokyo.

3 *Ibid.*

A *strategic knowledge community* is not exclusively restricted to the history and the spatial limits of an organization but is led by projects.

9.3. The Human Health Care (HHC) program

The pharmaceutical group Eisai implemented *knowledge creation* as a real management philosophy. This led it to tackle its very reason to exist as an enterprise and to redesign strategically its mission promoting the values of HHC as a compass for anyone inside the company. In doing so, Eisai succeeded in mobilizing more that just its own potential. Within the global project of *Human Health Care* for third-age persons, the group set up a *ba* that contributed to knowledge creation.

The HHC program becomes the chief orchestrator of a *strategic knowledge community* that merged collaborative efforts both from patients and from the company. Previously, Eisai had considered its own competencies and knowledge (in R&D, management, administration, business, etc.) as insufficient to achieve the goal of HHC for society at large. To undertake such a mission, other competencies and knowledge were required and they had to look within the environment of ill persons themselves: their families and their careers. Patients' understanding of their illnesses represented a major source of information and knowledge, so matching it with the capabilities of Eisai was a very strategic issue. Haughty, supreme, isolated and cut off, scientific knowledge was not sufficient. It had to be combined and articulated with the energy of demand through close interactions to give birth to what it was necessary to improve and to provide. In doing this, researchers from Eisai did not *privatize* the needs of patients, monopolizing them in the name of an indisputable hyper-competence, but they strived to contribute to solutions within a global movement.

The knowledge of ill persons, their relatives, nurses, etc. was above all tacit, silent and non-formalized. To create the appropriate conditions for externalizing this knowledge, a *ba* or *strategic knowledge community* brought together Eisai and patients within a common integrated effort (compass). The first step of the SECI model, empathy, provided the right climate for externalization of tacit knowledge that was not expressed in words or sentences. Eisai people involved themselves in practical attention to and care for patients. Through this relationship, people from Eisai perceived and felt in a tacit way difficulties, pains, apprehensions and rhythms:[4] "We are Ladies and Gentlemen serving other Ladies and Gentlemen."[5]

---

4 This kind of relationship sounds a little bit like *hara to hara communication* between the master (sensei) and the disciple in Japanese arts.

5 A slogan used for the HHC program within Eisai.

To maintain and to deepen the HHC philosophy, a formal smart grants system was set up to reward (in their salaries) those who contributed to improving and increasing the knowledge and the performance of the group. Listening, exchange and cross-fertilization between departments was greatly encouraged and also discussions about ways of thinking and processing. The fight against viscosity of information was ongoing. Files about persons met outside of the group were circulated and made available through the company's intranet. Anyone who looked at an item of information was able to give it an award point if he or she considered it worth it. Transverse training within the enterprise and benchmarking with firms that had the highest customer satisfaction were sought.

A *strategic knowledge community* functions on the basis of systematic flows and exchange of information that are very different from the *black hole ways* that absorb but do not allow any information to go out. To help doctors to improve their knowledge and practices, Eisai organized training sessions and made available an open and free call centre. People suffering from Alzheimer's disease in Japan also had access to a specific home page. Public lectures were planned and the group contributed to the annual day about this degenerative disease. All these practices acted in favor of the setting up of *ba* that brought together all those from outside Eisai who shared concerns about this illness and wished to increase their knowledge about it. Requests, opinions and comments coming from patients and carers in hospitals were processed with a lot of care. Questions without answers and diagnostics without solutions were of principal importance because they tackled and informed about *knowledge fronts*. The processing was detailed and the information was systematically distributed to enrich project management. Circulation went on until it led to qualitative and cognitive steps. Information and communication technologies accelerated the process.

Let us consider now the practical case of patients' fear about endoscopies, to which Eisai applied the four stages of the SECI model. The socialization step of tacit knowledge took place between doctors, researchers and psychologists through a questionnaire. Externalization of this knowledge made it possible to set up a second questionnaire for the patients to answer. Sessions to make them sensitive to the importance of endoscopies were planned and explanations and details were provided too. The results of both questionnaires were combined, analyzed and then communicated through new training sessions. The internalization step improved the doctors' knowledge and contributed to reducing patients' fears thanks to the development of their knowledge about endoscopies and the necessity for them.

As a result, making fears explicit and matching them with scientific knowledge gave rise to a new knowledge that enlightened both patients and doctors. This *knowledge movement* transformed the partners involved for their own global benefit. Efficiency and action came along with knowledge creation. This case study illustrates the importance of seeking to combine knowledge from diverse sources in order to move forward on *knowledge fronts* and to develop capabilities. What was unknown by the patients was not considered as an obstacle but of vital interest in solving the problem, in enriching knowledge and opening up possibilities. Everything had its importance and the experts did not consider themselves as the exclusive authorized owners of supreme and isolated knowledge!

9.4. Shaping a new way of functioning for organizations

Mutual respect and self-confidence played a major role and the knowledge creation process was based on dialectic relations. Doctors evaluated the fears *with* the patients and found a way *with them* throughout in order to provide a convenient method. Distinction between subject (active and who were supposed to know) and object (passive and who were supposed to be ignored) disappeared because of the management of a common project. An informational movement was led by a community effort to reach satisfactory solution. At the beginning convenient and operational knowledge was not available. Sharing the management of questions using the SECI model brought together scattered elements. Because the move was right, knowledge was created. The HHC program provided conditions of real *operational harmony* in the meaning given by John Boyd (see OODA Loop on www.belisarius.com). The HHC philosophy and values acted as a compass for everybody inside the firm for them to develop their activities and interactions. Sharing vision and project afforded a margin for maneuver in decision-making that resulted from the signal perceived by Eisai people, and functioned as a binding component that focused energy. Using the HHC standards, each person had the capacity to measure their own self-efficiency and performance as well as those of others.

Strategic knowledge communities (SKC) tend to loosen and dissolve the physical limits of organizations, fostering collaborative projects that include potentialities from outside. Partners and even competitors, customers and users, complementary sources of knowledge and competences, are engaged synergistically toward *operational knowledge fronts*. This dynamic porosity appears as a condition for strategic agility [BOY 04] within a hardly predictable knowledge-based society. As a result, time and rhythm in knowledge creation take the lead over space and physical limits that are too slow in terms of flows and processes and too limited as regards as the spectrum of a possible mobilization of resources in an opened-up fast global world.

The community principle is ruled by projects, aims and goals; it is therefore strategic in enlightening the immediate future. The key strategic principles of liberty of action and economy of means [FAY 00] can be applied within such SKC. In an open informational society [CAS 98] where information controls the raw material, the agent of transformation and the final product, the Japanese concept of *ba* offers the advantage of stressing the pre-eminence of relational frameworks so that information and communication technologies contribute in a dynamic way. The scope of experimentation is not just wide but imperative.

9.5. References

[BOY 04] BOYD John, *On creation and destruction*, www.belisarius.com.

[CAS 98] CASTELLS Manuel, *La société en réseaux*, Fayard, Paris 1998.

[FAY 00] FAYARD Pierre, *La maîtrise de l'interaction. L'information et la communication dans la stratégie*. Zéro Heure Editions Culturelles, Paris 2000.

[JAV 02] JAVARY Cyrille J.-D., FAURE Pierre, *Yi Jing. Le livre des changements*. Albin Michel, Paris 2002.

[JUL 02] JULLIEN François, interview in *Sciences Humaines*, no. 125, Paris, March 2002.

[NIS 90] NISHIDA Kitaro, *An Inquiry into the Good*, translated by M. Abe & C. Ives, Yale University Press, New Heaven, CT, 1990.

[NON 98] NONAKA Ikujiro, "The Concept of '*Ba*': Building a Foundation for Knowledge Creation", *California Management Review*, vol. 40, no. 3, Spring 1998.

9.6. Further reading

DAVENPORT Thomas H., PRUSAK Laurence, *Information Ecology. Mastering the Information and Knowledge Management*. Oxford University Press, New York and Oxford, 1997.

FADOK David S., *La paralysie stratégique par la puissance aérienne*, John Boyd and John Warden, ICS, FEDN, Armée de l'Air, 1998.

FAYARD Pierre, *Culture stratégique et technologies de l'interaction. La voie japonaise de la création du savoir*. Embassy report (France in Japan), April 2002, available on the website of the Agence pour le Développement de l'Information Technologique (www.adit.fr) and on www.stratego.tv.

FAYARD Pierre, *Comprendre et appliquer Sun Tzu. La pensée stratégique chinoise: une sagesse en action*. Dunod – Polia Editions, Paris 2004.

FAYARD Pierre, *Manager dans la société de la connaissance. L'exemple japonais*, Dunod, Paris, 2006.

FUMIKATSU Tokiwa, *Management based on Nature's Wisdom*, Tokyo 2001.

JULLIEN François, *Traité de l'efficacité*, Grasset, Paris 1996.

KIKAWADA Kazue, *Knowledge Dynamic Initiative*, Fuji Xerox, Tokyo, 2000.

MISHIMA Yukio, *Le Japon moderne et l'éthique samouraï*, Arcades Gallimard, Paris 1985.

MUSASHI Miyamoto, *Gorin-no-sho, Traité sur les cinq roues*, Maisonneuve & Larose, Paris, 1985.

NONAKA Ikujiro, *Managing Industrial Knowledge. Creation, Transfer and Utilisation*, (with Teece David), Sage Publications, London, Thousand Oaks and New Delhi, 2001.

SOCIETÉ Franco-Japonaise des Technologies Industrielles, *Les chemins du savoir au Japon*, Paris 2000.

TSUYUKI Emiko, *Collective creativity based on "Ba" theory*, available on www.stratego.tv (February 2002).

TSUYUKI Emiko, *Ba and Knowledge Creation*, PhD directed by Ikujiro Nonaka, Hitotsubashi University, Tokyo, 2003.

PART 3

Club Workshop Studies

Chapter 10

The Knowledge Maturity Model

10.1. Introduction

Over the last 20 years, a few companies have tried to launch programs dedicated to the management of their intellectual capital. Some of these experiments have been successful, while others have probably mobilized a lot of resources for a few results. However, the pioneering organizations obviously have little inclination to share their know-how and, therefore, Knowledge Management (KM) has remained a vague concept; it is remarkable that even consulting firms are still experiencing difficulties in turning it into a profitable business.

In the last few years, however, an increasing number of companies have started to understand better what it was all about. The internationalization of all economic spaces, the growing complexity level of industry and the fast innovative pace of some sectors are facts that automatically mean that increasing numbers of employees have to deal with dangerously fast growing volumes of information. IN parallel to this continuous and increasing learning, it appears more than ever necessary to preserve the knowledge acquired by the firm and therefore to transfer it from experienced staff to newcomers. And, of course, in this environment of growing competition, it is particularly recommendable that this process remains as fluid as possible.

It thus becomes clear that KM is very likely to have a strong impact on the economic results of an organization. In order to evaluate this, several organizations, essentially North American, have set up methods that allow the valuation, in an

---

Chapter written by Jean-François TENDRON.

accounting sense, of certain parts of the intellectual capital. Four of the better known methods are:

– *human resources accounting* (based on the analysis of the employees' costs and allowing the measurement of their contribution to the company results);

– *economic value added* (based on the accounting value of the net assets and allowing the optimization of the shareholders' interests and the optimization of the capital use);

– *the balanced scoreboard*, based on the tracking of four types of indicators – finance, customers, internal processes and learning – and allowing the identification of key success factors);

– *intellectual capital* (based on the tracking of three types of indicators – market, human and structural – and allowing the valuation of certain intangible resources).

Analysis of these sophisticated methods shows that they required important investments from the companies in terms of time, of training and of dedicated staff. Furthermore, they do not reveal the positioning of the companies according to the state of the art in KM (the Intellectual Capital method is probably the most satisfying one in this respect but the elasticity of the bounds between the market, human and structural capitals make the intercompany comparisons virtually impossible).

Therefore, there is a need for a starting point to this reflection. A starting point that will help realize what it is possible to achieve in terms of KM. Before going any further, candidate companies need simple tools to evaluate their own capacities in the matter.

10.2. Work methodology of the commission on "Aspects économiques de la gestion des connaissances" (economic aspects of KM)

KM depends on numerous factors: economic, cultural, managerial, technological, methodological, etc. And all organizations already manage certain parts of their knowledge capital in a manner that is more or less aware. It is therefore difficult to represent the capacity of an organization in terms of KM without utilizing a point of reference that can at least describe the existing situation.

Consequently, the commission on "Aspects économiques de la gestion des connaissances" decided to produce a grid of criteria allowing the evaluation of the knowledge maturity of an organization.

The elaboration of this grid was based on previous research led by Microsoft and Thalès.

Microsoft has developed, in collaboration with other enterprises (Delfin Systems, Ernst & Young, PriceWaterhouseCoopers, Compaq), a tool called "IT Advisor for KM", which is aimed at helping its clients in the evaluation of their KM best practices. The tool is a grid of criteria scaled from 0 to 3 and covering the three following topics: the value of know-how (people), the value of products, systems and services (process and technology) and the value of clients and partnerships (business relationship management).

In 1999, Thalès (Thomson CSF) launched a workgroup in collaboration with Raytheon, Nortel, Sun Microsystems, the CEA (French agency for nuclear energy) and Twente University. This group has produced a grid of criteria scaled from 0 to 3 and allowing the evaluation of the capacity of an organization in terms of KM. The criteria of the grid are grouped into four categories: knowledge structure, collaborations, KM tools and methods.

Based on these two research programs, the commission proceeded in an iterative way in order to progressively integrate the experiences of the different club members (Figure 10.1).

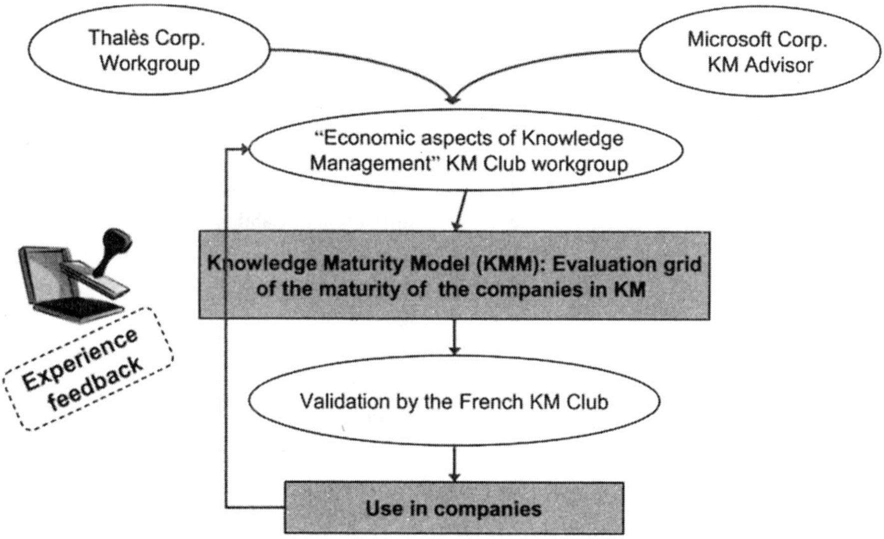

Figure 10.1. *Principal functioning of the KM Club workgroup*

A few guidelines were set at the beginning. For example, the selected criteria were to take into account the tacit and explicit dimensions of knowledge as well as the incitement to participate in knowledge consolidation. The grid was also supposed to show the organizational dimension of KM.

10.3. The Knowledge Maturity Model (KMM)

Version 1.0 of KMM was produced in October 2001. This grid is neither normative nor exhaustive and it may look a bit disconnected from the initial objective of the commission, which was to evaluate the economic aspects of KM. Nonetheless, it does initialize a methodology that has become necessary in order to approach KM issues in enterprises.

Version 1.0 of KMM contains 18 criteria grouped in seven different topics (Table 10.1). All the criteria can be measured on a scale of four levels (from 0 to 3). In order to facilitate the comprehension of the grid, the significance of each criterion is developed in the form of one or two questions and all levels are described succinctly.

Topic axis	Criteria	
Information processes	1.	Information formatting
	2.	Information acquisition modes
	3.	Information validation
	4.	Information accessibility
	5.	Information reusability
Tools	6.	Categories of tools
	7.	Information security
Methodologies	8.	Training
	9.	Processes of knowledge consolidation
	10.	Production
Collaborations	11.	Interfaces
Management	12.	Engagement of top managers
	13.	Motivation
	14.	Competency management
Culture	15.	Comportment
	16.	Conviviality
	17.	Identity
Economic aspects	18.	Intellectual capital

Table 10.1. *The 18 criteria and their groupings*

THEME : INFORMATION PROCESSES
Criterion : Information Reusability

Which is the effective (productive) level of information re-use in the company?

Level 1: Capitalization

Accessible databases on-line contain information on existing devices, modules and products. The level of consultation is low. No significant case of re-use can be shown. *A priori*, the elements of the base are not designed to be re-used.

Level 2: Transfer

The level of consultation of information bases is important and regular. We can identify some significant cases of profits per recovery and/or adaptation of elements of the base. Some of these elements result from a produced policy or module of the company.

Level 3: Appropriation and Enrichment

A culture starts to be established in which the personnel re-uses/adapts/modifies then enriches the base with new elements. More energy is spent to improve rather than to rediscover (Reduction of the NIH – Not Invented Here). The elements are capitalized in a better adapted form to the re-use.

Level 4: Creation and Innovation

At the time of the stages of definition/design, the systematic consultation of already existing solutions, devices and modules can start a process of creation and innovation by association and assembly. It can thus result from a competing advantage for the company and/or a reduction from the time-to-market.

Figure 10.2. *Example of a criterion*

Even by itself, without any rules for its use, the KMM grid already represents an interesting product because reading it raises questions that do not come naturally.

For example, it is remarkable that, during the implementation of knowledge portals, most managers often tend to remain focused on the role of technologies, which is a mistake. Indeed, a simple reading of the KMM shows that motivation, identification with the project and conviviality are key success factors.

In the field of KM, the notion of risk is well illustrated by the observation of the differences between top-down and bottom-up projects: the risk is inversely proportional to how much the organizational dimension of KM is taken into account.

An example of typical bottom-up project was initiated during the 1990s by a well known European aircraft manufacturer. A small team of engineers launched a local project that allowed, within a few years, the automatic configuration of space in commercial aircraft. The organization carrying out the project built itself very progressively, thus acquiring an important level of maturity regarding KM issues.

In comparison, the success of top-down projects is often more random. Also in the 1990s, many consulting firms tried to provide their consultants with tools designed to access the knowledge bases of the whole organization. The technology was practically never an obstacle, but the requirements in term of change management were often so huge that they damaged the project feasibility.

By definition, top-down projects primarily aim at strategic targets, thus placing the organization in a situation that is sometimes difficult to apprehend. The approach often results in a technological solution that is incompatible with the multiplicity of points of view required by the KM actions.

In this context, the KMM is a sort of checklist. The taking into account of most of its criteria allows an important reduction of the risks included, particularly in the top-down configurations.

10.4. Use of the KMM

We have, so far, experimented with two types of KMM use.

10.4.1. *Raising consciousness about KM*

The first use aims to increase the awareness of a group about KM. An administration concerned with child benefits ("C.A.F.") and a small company specializing in electrical installation gave over a day to a directed reading of the KMM. In both cases, external consultants were there to explain the significance of the different criteria, as well as to direct the debate on the basis of their experience in the field of KM. The outcome was very positive since the exercise brought the leading committees to ask themselves questions about topics that would not have arisen naturally, such as the direction of engagement in the field of KM. (For C.A.F., the exercise was followed by a mapping of the organization's critical knowledge.)

The principal interest of KMM, within the scope of a consciousness-raising workshop, is thus to provide what Americans are used to calling "situation awareness". Probably for cultural reasons, this expression seems quite inappropriate in a French context. It is, however, the best way to describe the impact of KMM. A reading of its criteria obliges one to take into account the factors that influence the capabilities of an organization in terms of KM.

The criterion "Evaluation of the intellectual capital" well describes the state of awareness of the company in relation to KM (Figure 10.3).

THEME : ECONOMIC ASPECTS
Criterion : Intellectual capital

Is there in the company an effective awareness of the intellectual capital?

Level 1: No awareness
No effective awareness of the intellectual capital in the evaluation of the company.

Level 2: Standard evaluation
The intellectual capital of the company is taken into account by the standard intermediaries of actions: patent, publications, quotations, press review, image, participation in mental activities external with the company (training, conference, committee, etc.). No rational evaluation of this capital.

Level 3: Specific evaluation
A standard evaluation exists. Moreover, specific and localized evaluations exist for certain programs, certain businesses or within the framework of certain operations (technology transfer, sale of license, etc.).

Level 4: Intellectual capital management
There is an effective and total intellectual capital management in the form of criteria, dashboard, financial evaluations. These elements of evaluations appear in the management reports as well as the financial reports.

Figure 10.3. *"Awareness of the company"*

It must be make clear, however, that in such a type of use, the KMM does not enable an organization to objectively evaluate its capacities in terms of KM. There are several obstacles to this. First, working at the level of a committee of top managers necessarily implies that the persons involved in the reflection have very different preoccupations. Thus, the sectors they are in charge of (human resources, production, accounting, etc.) may have very different levels of knowledge maturity, so that the average of the levels obtained from the criteria does not mean anything.

Furthermore, the self-evaluation that forms part of the KMM is quite difficult. Even if we suppose that the candidate remains perfectly objective, there is still an obstacle that comes from the way the model has been built. As consensus was the rule when the KMM was being built, it was often necessary to simplify the description of the different criteria, so that the expertise concentrated in their definition does not appear in an explicit way. Consequently, a candidate for self-evaluation should be aware of what is achievable in terms of KM, which is practically never the case.

10.4.2. *Evaluation of a community's maturity in terms of KM*

A second type of use aims to evaluate the capacity of a community in terms of KM. A whole organization may have many targets, each very different from the others. In contrast, a community (of interest, of practice) usually aims at a single objective.

In order to avoid the risks inherent in any sort of self-evaluation based on the KMM, we decided to collect the maturity-related information by interviewing persons representative of the group.

Thanks to the presupposed homogeneity of different people's objectives, the average level of maturity for the criteria then really means something.

10.4.2.1. *Example of evaluation*

The example below (Figure 10.4) shows the results of a survey on a community of interest. All the persons interviewed work on the development and the marketing of an economic intelligence product. The company digitizes the content of more than 700 publications a month and sends to each of its customers the articles corresponding to pre-determined keywords. These articles are received and stored using software developed and provided by the surveyed company. On demand, a re-examination service can substantially increase the level of detail of the articles.

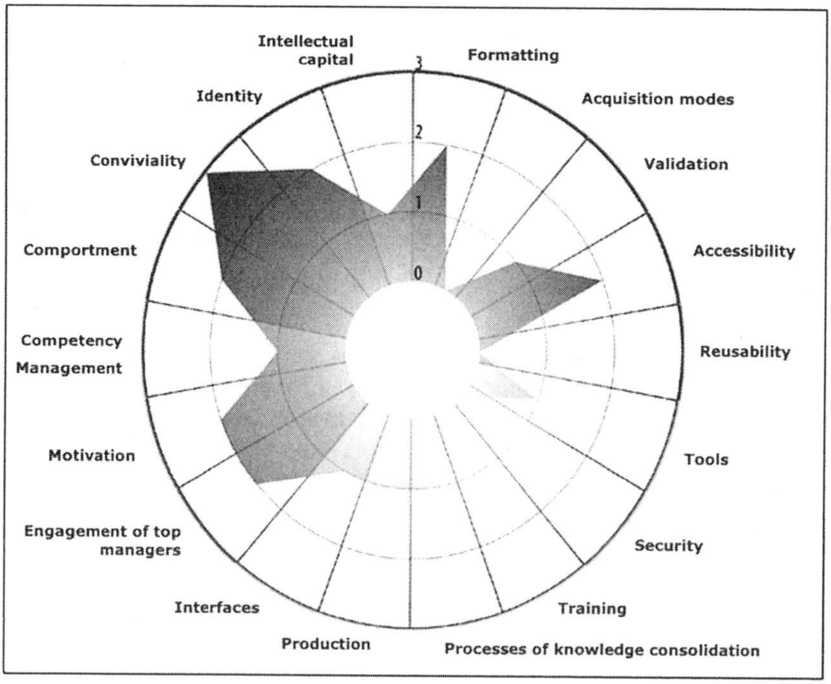

Figure 10.4. *Maturity of an economic intelligence service (community of interest)*

The assessment is quite surprising. The staff of the surveyed community put at the disposal of its customer a tool that performs. The paradox is that they do not seem to have questioned themselves about their own mode of knowledge acquisition, nor about their capital of potentially reusable knowledge. That situation may be explained by the fact that most of the reflection concerning internal KM has been left to the IT service (it must be admitted that it has effectively carried out its duty: most of the information is potentially accessible).

We can also see that in production (digitization, article selection, information reexamination, development of the customer-dedicated software), most of the staff have to search themselves for the information they need. Consequently, the consolidation of knowledge is essentially opportunist or conjectural; there is no planning in that sense.

Finally, the KMM shows that the enterprise culture is potentially very compliant with future developments in terms of KM. In particular, the community being quite young, the staff do not feel any problem when they communicate in transverse ways or at all hierarchical levels. Even if this communication essentially remains opportunistic at the time of the diagnostic, it still represents an important asset: as a matter of fact, the cultural obstacles are the most difficult to get rid of.

This use of the KMM required quite an acceptable time: 10 interviews of two hours each were enough to build a reliable maturity profile regarding KM for a community of 50 persons.

10.4.2.2. *Example of reactions*

One of the principal interests of such a diagnostic is to make a priority the actions to be carried out in order to make a community a little bit more of a "learning community". In the case of the former example, it is obvious that a workshop of consciousness raising about information acquisition is a prerequisite to any action. It must be demonstrated to the staff that, by giving more sense and more context to their knowledge contribution, they will really contribute to increasing the collective intelligence.

In fact, the people in charge of the customer satisfaction assessment are making inquiries about the software delivered and transmitting topics to the IT team. However, those surveys are left to the entirely free interpretation of the computer people. Consequently, and whilst all the information necessary to anticipate customers' needs is theoretically available, the changes between different software versions are only the result of technical compatibility issues.

The second priority, in the case of our example, is to force the community to question itself about the part of reusable knowledge it is producing. All the interviewed persons punctuate their learning with bibliographies of tricks or best practices, often stored on classical "bureautic" tools (MS Office), but this know-how does not benefit to the whole community. The company's memory is thus practically unexplored.

In the case of a complex industrial organization, analysis of the knowledge acquisition modes would probably require a preliminary mapping of the critical knowledge in order to isolate the important contents and to guarantee an acceptable return on KM processes. For a community of 50 persons working in a relatively simple business, however, the intervention of an internal or external consultant is generally sufficient to help the whole group progress on these two topics.

10.5. Perspectives

The problem is that it is difficult to plan the intervention of KM specialists (internal or external) because there are very few of them. Thus it is likely to become one of the biggest challenges of the next few years: to train qualified personnel in the field of KM.

Nevertheless, this field remains emergent and the required competencies seem at the same time so vast and vague that it is difficult to determine them. One thing, though, is clear: it is urgent to reverse the tendency according to which IT specialists, having a technological point of view, should be the principal persons in charge of KM in organizations.

From such a perspective, the KMM can be used as a training support for the future KM staff of an organization (Figure 10.5).

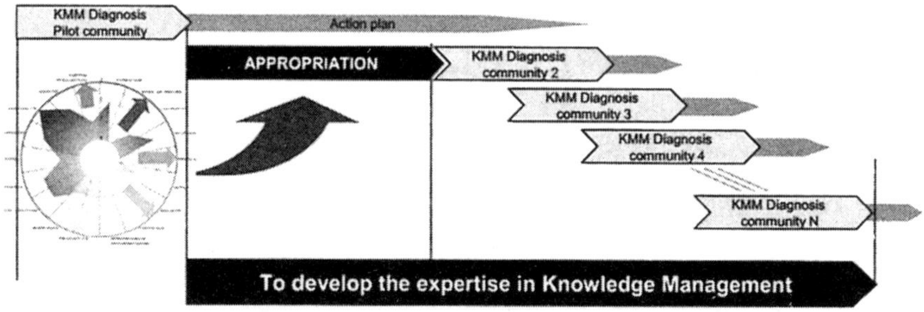

Figure 10.5. *Use of KMM for internal expertise development in KM*

As the first activity of a pilot community it is necessary to customize the KMM criteria to the language of the organization, (if possible, within the limits of the existing framework). Second, quick training should make it possible for the staff who are likely to be involved to integrate both the KMM as and how it is used.

These people can then diagnose the maturity of other communities in the organization. Their objective is to determine the actions required by weaknesses demonstrated by the KMM, as well as to complete their training with a practical exercise. The time needed for a serious diagnostic is such that, whatever the community considered, it obviously does not mean a full-time job. This is an important issue because the organization must be able to test different types of candidates (there is no existing profile for a knowledge manager).

A training program of this type can then provide the organization with an internal network of KM specialists. This approach is probably one of the first realistic steps toward the "learning organization".

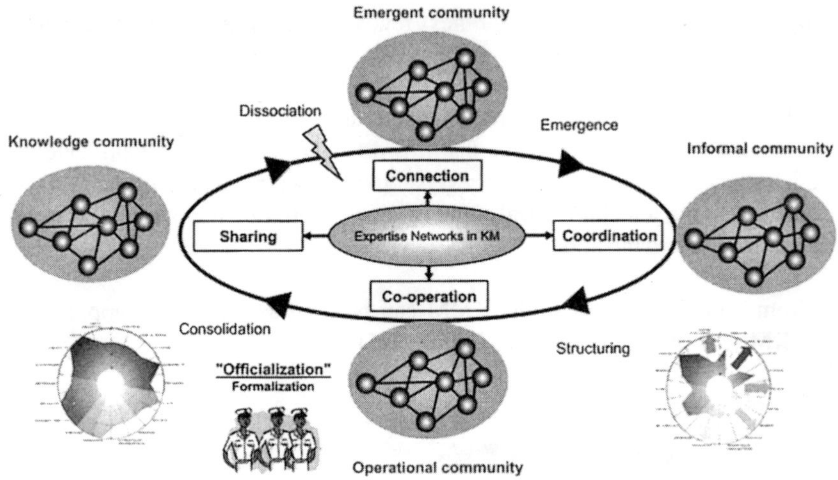

Figure 10.6. *Expertise networks in KM*

The internal network of KM expertise should enable the organization to create learning communities relating to its main objectives. In such a case, the concept is no longer to reinforce existing communities but simply to create them according to the corporate targets (Figure 10.6).

The obvious interest of creating knowledge communities according to a given objective is to rapidly generate transverse flows within the organization. Experience shows every day that this is far from natural behavior for most of the people concerned. From that perspective, the length of life of a community can sometimes be of no importance; it is the capacity to create communities that forms the strength of the organization.

10.6. Conclusion

The work of the French KM Club ("Club Gestion des connaissances") has made it possible to build a framework that allows the positioning of any company according to the state of the art in the field of KM: the KMM.

This framework can obviously be improved and it probably requires further development according to the targeted organizations. It is, for example, very likely that extended enterprises will add criteria to the "collaboration" topic. One can imagine completing or transforming the KMM in an "Extended enterprise maturity model". Such a development could occur within a relatively short time. Companies facing complex supply chains are often pioneers in the field of KM. As a matter of fact, it is much easier to exchange computerized information on flows than it is to computerize the flows themselves.

Whilst allowing a reasoned evaluation of the risks prior to a KM action, the KMM should soon provide a good basis for benchmarking work. And we think that, in the future, only benchmarking can provide the value of the intellectual capital of organizations that is based on reality.

In that way, the KMM is an obligatory checkpoint toward taking into account the economic aspects of KM.

10.7. References

[CEFRIO 03] Centre Francophone de Recherche en Informatisation des Organisations, Quebec, lecture organized by *Information Highways Magazine*, "Les défis de la gestion des connaissances", February 2003.

[Club 00] Club Gestion des Connaissances, Dorothée Touboul, *Méthodes d'évaluation de l'immatériel et problématique de l'évaluation économique des connaissances*.

[Club 00] Club Gestion des Connaissances, J.-L. Ermine, H. Laude, *A Knowledge Maturity Model*.

[Club 01] Club Gestion des Connaissances, *Modèle de maturité cognitive ou Knowledge Maturity Model version 1.0*, October 2001.

[Ermine 02] Ermine Jean-Louis, Initiation à la méthode MASK, *Le processus de développement des communautés de savoir*, CD-ROM edited by the Technical University of Troyes, 2002.

Chapter 11

Knowledge Mapping: A Strategic Entry Point to Knowledge Management

11.1. Why map corporate knowledge?

Companies assess the risks to which they are subjected in order to determine their strategy. *Major risks,* in other words those that are likely to endanger the company, are identified, mapped and assessed first and foremost.

Risks of loss of knowledge and crucial know-how and loss of development opportunities are increasingly considered as *major risks* by companies, whose intangible capital represent a prime resource. Specifically, companies prepare to face a surge of retirements and therefore to confront major risks of losing knowledge and know-how.

Similarly, companies identify, map and assess their *competency capital.* Corporate collective competencies are specifically linked to *transverse processes,* on which *quality approaches* are based. One characteristic of these competencies lies in the fact that they are difficult to imitate and highly dependent on the specific company context [Tarondeau, 98].

Chapter written by Gérard AUBERTIN, INRS.

Companies have developed approaches that may be direct (Knowledge Management (KM), competency frameworks) or indirect (quality approach), but are invariably aimed at preserving and developing their intangible capital.

Nevertheless, it is observed that:

– knowledge/competency management approaches are often uncorrelated with respect to corporate strategy;

– competency and KM approaches are insufficiently coordinated; KM approaches are often directly managed by business-related line managers as and when departures arise; competencies are often managed by Human Resources departments with sometimes insufficiently reactive consideration for business development;

– quality approaches give insufficient or over-formal consideration to KM;

– uses to which competency and core competency frameworks are put are often ambiguous in relation to personal assessment;

– knowledge, competency and quality management tools are used independently of each other.

Typically, integration of these different approaches provides added value to the company. Mapping tools (knowledge, competencies, processes) coupled with criticality studies facilitate this integration. Moreover, these approaches represent an interesting entry point to knowledge/competency management at the highest level, through involvement of company general management in these issues.

Figure 11.1 shows in diagrammatic form, the links between different types of mapping, frameworks and associated management systems. In particular, this figure shows that competency and core competency frameworks, along with associated management systems, in fact often respond to a twin aim of critical knowledge/competency management and personal assessment.

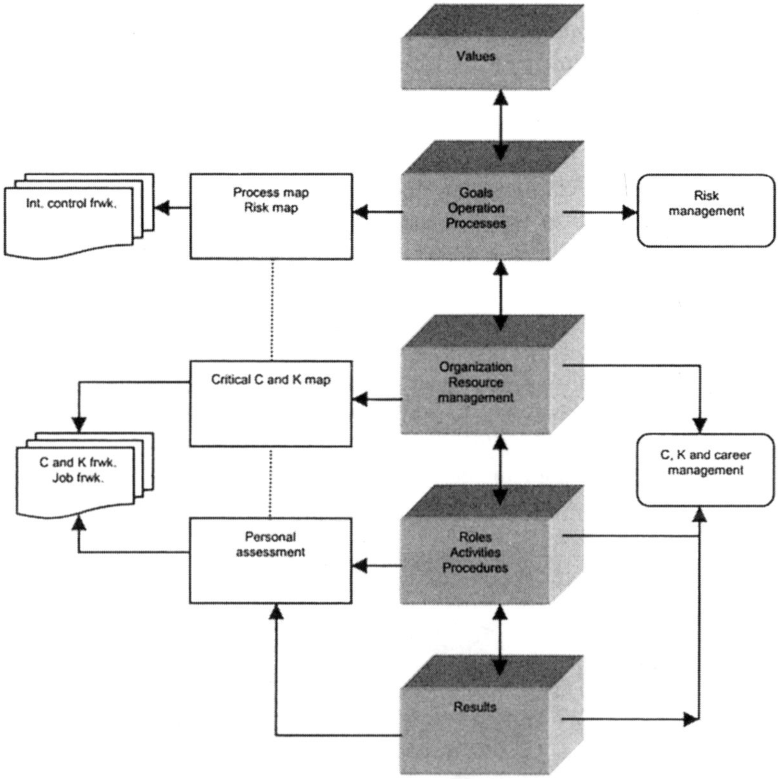

Figure 11.1. *Links between types of mapping, frameworks and management systems (C = competencies; K = knowledge; map = mapping: frwk = framework)*

11.2. What knowledge and competencies should be mapped?

Knowledge and competency concepts are often perceived as magical, as "strange attractors" that focus company attention and efforts to improve their performance. However, these concepts remain vague and include multiple aspects and practices, depending on the company. It is not our intention to conduct a review of the subject, but to introduce briefly the meaning we have assigned to these concepts, within the scope of this project.

Knowledge is sometimes defined as information that is meaningful in a particular context and for a given person, because it is so much linked to personal representations. *Knowledge* is primarily made up of *theoretical knowledge* and *methodological knowledge*, in other words *structured knowledge* operating at a personal level. *Competencies* are the capacity for mobilizing, combining and integrating knowledge into action with an end-purpose in mind. They are based on knowledge, which they integrate, as well as on little formalized (indeed difficult to formalize) *know-how and practical knowledge* required for their implementation (Figure 11.2). Areas of knowledge therefore constitute the structured part of competencies. Knowledge can also be considered as potential, unfinalized competency. It becomes full competency when it is mobilized in the form of activities and processes. Integration and making explicit these competencies in a person's framework leads to enrichment of the knowledge possessed by that person.

Competencies have been defined here in their strictly *cognitive dimension* and this means that the *behavioral dimension*, which is often hidden behind the concept of *know-being,* has been excluded. Such an approach is similar to the one that states that a competency is "that which enables us to define the way to acquire it" [Lichtenberger, 97]. According to this design, this dimension of knowledge would be better taken into account in the concept of *role*, defined in core competency frameworks, than in the concept of "know-being" [Penso-Latouche, 00].

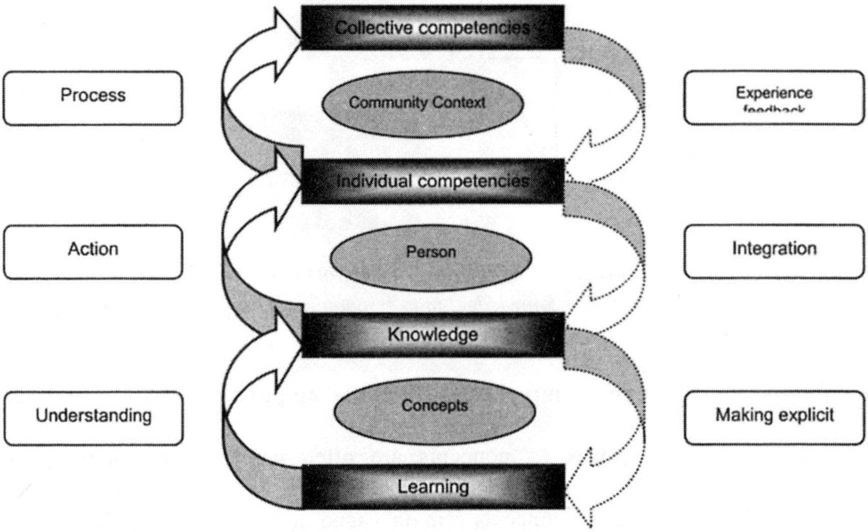

Figure 11.2. *Competencies and knowledge*

In practice, two approaches to managing competencies are commonly used by companies: the "business-related competencies" approach (know-how and professional expertise) and the "generic or managerial competencies" approach (intra-personnel, inter-personnel, cultural, etc. competencies) [Geffroy, 02].

The first approach focuses on developing both business and key core competencies and (technological) resources. Dominant factors are core competencies, professionalism and expertise. This approach is often developed in companies with a strong engineering culture and involves mainly technicians, researchers and non-executive personnel, with a view to enhancing collective performance. The "competency" system is highly structured, most frequently according to the line of work. This approach corresponds to the one developed within the present project scope.

The second approach focuses on individual performance, results and innovation. The aim of "generic competency" management is to manage individually the competencies and careers of managers and experts, especially "high potential" people ("high flyers"). In part, it is based on the hazy notion of *know-being,* combining the cognitive and behavioral dimensions associated with this concept. These systems concern principally executive personnel. Inter-job mobility is sought; dynamism and managerial capacities are extolled values. They presuppose implicitly or explicitly *expected behaviors* and *involvement* on the part of people. "Generic competencies" are the opposite of technical–functional competencies. In principle, they would be better managed using the concept of *role* than that of competency. They have not been considered in this project.

Knowledge Management

The problem of KM is conventionally defined as the setting up of a cognitive flow management system, which allows all organizational components to both use and enrich the organization's knowledge capital. KM sets out to *identify, share, enrich and enhance the status* of company knowledge and especially knowledge of a particularly critical nature. This is undertaken with the aim of promoting the capacity for innovation and creation by increasing productivity of knowledge and know-how.

Setting up a strategic plan for managing knowledge implies replying to the following four key questions:

- What knowledge forms part of the company capital?
- What knowledge is critical to this capital?
- What critical knowledge has not been mastered?
- How should use of this knowledge be optimized (risk reduction)?

Company cases considered

This chapter presents the results of knowledge/competency mapping operations conducted at seven companies, all members of the "Club Gestion des Connaissances" (CGC) (Knowledge Management Club). Specifically, it relates the experience acquired within INRS, an information, training and research organization working in the occupational risk (health and safety at work) field, which initiated the thought process undertaken within the CGC.

11.3. How is knowledge/competency mapping performed?

11.3.1. *Defining the mapping goal*

Performing mapping is not an end in itself. What are the end-purposes of the approach? Is the aim of mapping to:

– facilitate informal exchanges and contacts between members of the company to encourage transverse processes (who does what?)?;

– detect knowledge available within the company? (who knows what?);

– identify working groups and existing communities of practice (who exchanges what?)?;

– facilitate emergence of communities of practice?;

– serve as a basis for a knowledge criticality study?;

– map processes and activities of agents?;

– recognize the role of each member of the company?;

– facilitate access to documents, information?;

– etc.

Such questioning allows the mapping approach to be positioned with respect to other adopted transverse approaches and to ensure the necessary links from the start. Issues faced in quality and competency management are in fact similar to those of KM, in that they attempt to create a working environment that enhances the productivity of knowledge and know-how.

Moreover, acquaintance should be made with the resources offered by the company information system, not only to assist in implementing the approach, but also to physically safeguard and encourage it.

11.3.2. *Identifying knowledge*

The first stage involves identifying the cognitive resources of the company or organization. Three types of approach can be distinguished:

– the functional approach, usually based on an organization chart;

– the procedural approach, based on company processes;

– the conceptual or area-based approach, organizing information around subjects, objects or end-purposes.

Characteristics of these three approaches are given in Table 11.1.

Type of approach	Characteristics
o Functional approach 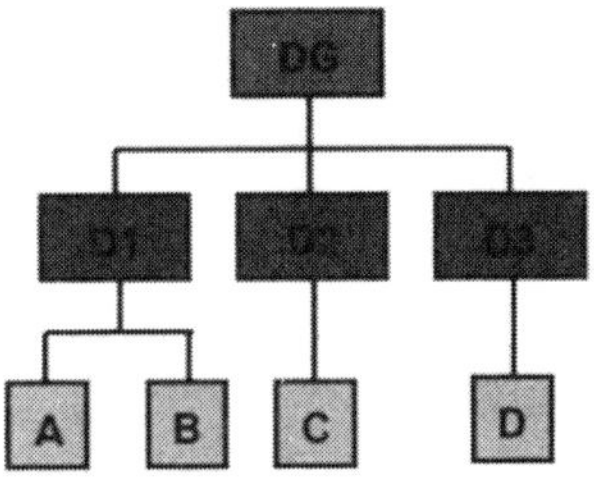	o Easy to implement o Flexible o Compartmentalized o Structure-dependent
o Procedural or process-based approach 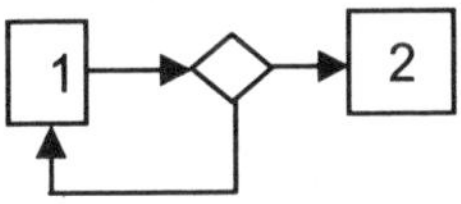	o Organization-independent o Based on added value o Requires defined processes o Relatively complex to implement
o Conceptual or area-based approach 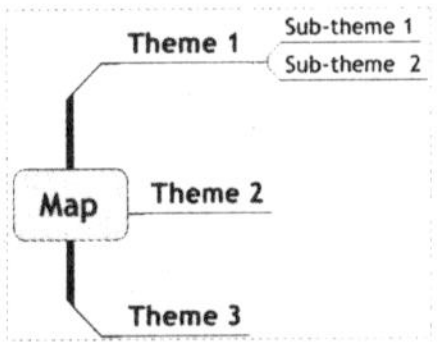	o Organization-independent o Based directly on resources o Suited to complex knowledge o Facilitated by in-depth knowledge of fields

Table 11.1. *Different mapping approaches*

The more the level of knowledge complexity increases, the more conceptual area-based organization systems tend to become the appropriate knowledge identification tool. They are especially well suited to technical systems such as research centers or business-based organizations. Conversely, organizations more directed towards production or support tend to adopt process-based approaches.

11.3.2.1. *Conceptual approach to identifying knowledge*

Identification of areas of knowledge using the conceptual approach involves highlighting such areas by successive analysis of corporate units and their activities, projects, products, etc. using reference documents and interviews.

Here, we are seeking to perform an "ascending" analysis, starting from a mass of information, with the aim of reorganizing this mass on a non-functional logic basis. This task requires high analyzing capacity. In this case, preliminary knowledge of company activities represents both an advantage and a handicap; an advantage in that preliminary knowledge of activities facilitates grouping of activities into coherent themes; a handicap in that knowledge of the company operational structure tends to influence thematic groupings. The approach comprises leaving aside the company operational structure by grouping activities into areas of knowledge and is somewhat unnatural in itself. It will be all the more arduous because, by operating habit, we tend to link such-and-such an activity with such-and-such a service or department.

Essential reference documents include the following.

– organization documents (Articles of Incorporation, organization chart, description of department activity distribution, directory of personnel activity, annual report, etc.);

– strategic documents (strategic plan, needs studies, summaries, etc.);

– production-related documents (customer enquiries, products, publications, studies, annual reports, etc.);

– quality documents.

11.3.2.2. *Process-based approach to identifying knowledge*

The approach is based on analyzing resources in terms of the knowledge that contributes to the smooth running of organizational processes. The proposed method is inspired by the GAMETH® approach [Gründstein, 00]. The implemented approach features the following three stages.

11.3.2.2.1. Stage 1

During the *first stage,* sensitive processes are identified in conjunction with those taking part. *Sensitive processes are processes that represent collectively recognized issues at stake.* Wanting to analyze all processes in a relatively short time is indeed difficult to imagine. Furthermore, analyzing a process may turn out to be out of proportion to the anticipated benefit. The method is therefore based on a heuristic approach that enables sensitive processes to be determined.

11.3.2.2.2. Stage 2

The *second stage* involves both *modeling* the sensitive processes retained and identifying activities and people associated with each sensitive process. This approach prevents dissociation of the knowledge possessed by people appointed within the company processes, of the actions they perform, of the decisions they make and of the relationships between them and their environment.

11.3.2.2.3. Stage 3

The *third stage* involves *identifying* the knowledge required to solve activity-related problems along with the mastered knowledge essential to activity operation.

11.3.2.3. *Identifying knowledge using automatic mapping tools*

Automatic mapping tools can provide assistance by facilitating the identification of areas of knowledge. However, the mappings derived *lack intrinsic meaning* and can therefore turn out to be deceptive and not fully comprehensible. They should be used with care and with a critical attitude.

11.3.3. *Building the areas of the knowledge map*

This section deals with the physical representation of knowledge mapping. The previous stage permitted in-depth work to be performed on company activities. We now proceed to finalize presentation of this analysis work with the important aim of making it accessible and usable. Representation must therefore be adapted to the operational vision of the people concerned with its use.

Knowledge can be considered according to different "viewpoints", each representing the way in which people see their activity. There are two major categories representing this knowledge:

– *"business related object"* representation: in industry, manufactured products, production technologies and possibly product design methods will be naturally identified; in a research center, the scientific fields covered will naturally appear;

– *systemic* representation, for example according to the well-known quality method featuring *methods, environment, resources, labor and material.*

This representation duality can be illustrated by representing INRS's health and safety knowledge. From the research center's standpoint, business-related knowledge was "seen" according to the "business related object" approach, in other words according to a description similar to the way in which researchers conduct their studies and generate knowledge. On the other hand, this same knowledge "seen" by people in charge of helping external inquirers was presented according to a systemic approach (see Figure 11.3).

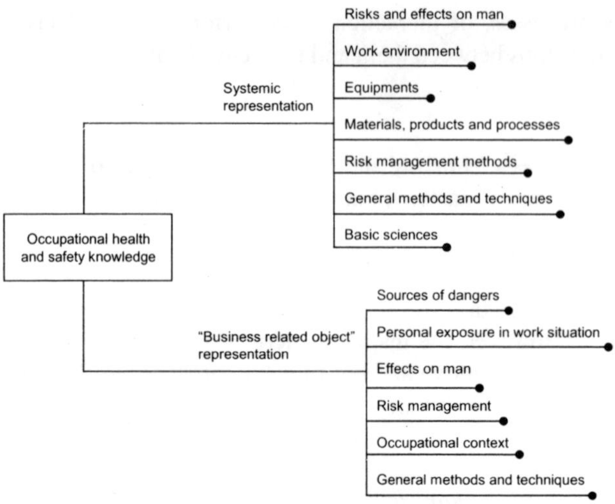

Figure 11.3. *Knowledge representation according to two viewpoints*

Map building is an iterative process and validation is performed by line managers as the process progresses. These managers are objective enough to position areas of knowledge within the organization and know the functional needs. Above all, they adjust the representation and appropriate it as interviewing progresses.

Constant three-step shuttling is in fact performed involving:

– problem appropriation;

– representation collective build-up;

– map validation.

Mapping can be extended to and integrated into knowledge, competencies and processes, to take into account different KM approaches. INRS-developed mapping was performed with this principle in mind. Three major axes were thus determined: knowledge of occupational health and safety, INRS know-how in exploiting occupational health and safety knowledge and INRS skills in relation to its operation.

11.3.4. *Representing knowledge: the area-based mapping model*

11.3.4.1. *Formal model*

The formal model, structured in the form of an UML-type class diagram (Figure 11.4) is a hierarchical representation, allowing company areas of knowledge to be classified into several levels:

– A *knowledge area* can be defined as the field of activity of a group of people, for which information and knowledge can be consolidated.

– The mapping central point is the *core business* or "core business knowledge", which corresponds to knowledge making up the strategic capital of the organization's body of knowledge and know-how in relation to its core business.

– *Axes of knowledge* emanate from this central point, defining strategic areas of knowledge, which often correspond to the organization's business operations.

In the classification, final areas of knowledge are grouped together along the axes of knowledge according to a common end-purpose or a similar knowledge theme. Depending on the level of accuracy of the area break-down, one area can be divided into sub-areas and one theme into sub-themes. Areas of knowledge can and must be linked to a group of people who possess structured or semi-structured knowledge and information. The model can thus be enriched with information models (document references, "Best Practices", etc.), which is particularly helpful when setting up a knowledge server.

A distinction is made between axes of knowledge that are directly linked to the core business and those concerning other cross-company areas. This means that the end-purpose of certain areas is not to respond to core business problems, although it is the case, for example, for knowledge linked to support activities (computing department, Human Resources, etc.). In this case, these *cross-company axes* are indirectly linked to the main activity.

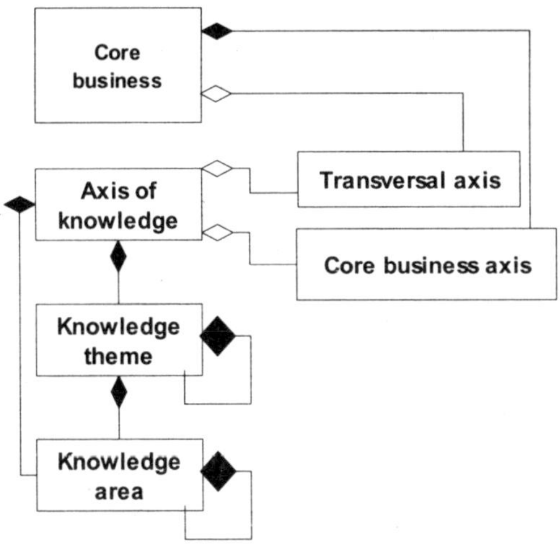

Figure 11.4. *Formal mapping model*

11.3.4.2. *The graphical model*

Graphical representation of knowledge mapping can be considered as a viewing method that facilitates navigation and provides an overall view of the company's areas of knowledge. Choice of a *Hishikawa*-type representation, for example, allows the hierarchy of the different levels to be presented in the form of spikes extending from a common trunk. Representation in the form of a *tree* is easier to implement because it allows graphical representation software to be used (Figure 11.5).

Representation of all break-down levels on a single map makes reading difficult. This is why it is preferable to represent mapping on several maps to facilitate legibility if multiple areas of knowledge are involved. A global view of the organization's activities and business areas is obtained from the most macroscopic levels. Representation going from the axes to lower levels provides a more accurate view of the knowledge area tree structure (six levels in INRS's case).

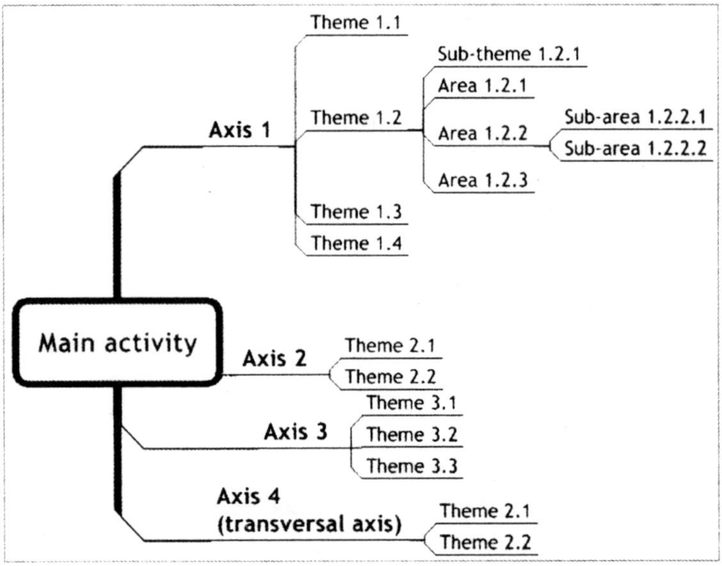

Figure 11.5. *Graphical mapping model*

In addition to hierarchical break-down of knowledge capital, graphical representation can be enriched by additional semantics: the larger the area (usually in terms of personnel numbers), the closer it comes to the center of the map (core business). Diagram line width can also vary according to the importance of the area.

Commonly, area positioning becomes more distant from the center as an area becomes more specific (less general).

Knowledge area mapping is essentially independent of organizational structure (organization chart). Knowledge can be shared amongst several units and people possessing knowledge in a common area do not necessarily belong to the same organizational entity (service, department, etc.). The building principle highlights areas of knowledge that are common to different services without referring to the organization structure.

11.4. What are the operational uses and mapping tools?

Using mapping entails foreseeing implementation of operational applications, especially computer-based tools for managing knowledge. Company knowledge mapping in fact constitutes decision-making and knowledge-sharing support, which can be more efficient in an electronic form.

Here, we address mapping utilization options and characteristics of tools that have been prototyped within the scope of the INRS-based application. The first use is presented in greater detail in the remainder of the text:

– *Representation of knowledge mapping and identification of people, groups and communities of practice.* This involves putting in graphical form data concerning areas of knowledge, to facilitate their access. The created representation is used as a static map background for other data. The map allows *identification of people, groups and communities of practice* within the organization and direct access to associated information. Furthermore, additional *statistical processing* provides useful analysis for managing competencies and jobs.

– *Representation of mapping and identification of critical areas of knowledge.* This function displays *criticality data* associated with areas of knowledge. Representation can, for example, appear in the form of a transparency laid over the mapping representation. The criticality of each knowledge area is represented by a color code based on the average assessment value. Three criticality levels can be distinguished, corresponding to the three sets of areas of knowledge revealed in the results analysis. Moreover, each area features a representation of the associated criticality diagram.

– *Representation of mapping and access to sources of knowledge associated with the area.* Areas of knowledge appearing in this representation serve as links to the different sources of existing knowledge. The source access interface should not impose specific technical structures on the implementation. Some sources can have easily been made accessible within the development framework. Indeed, this is the case for company bibliographical databases and its directory.

11.5. Knowledge mapping and criticality study

11.5.1. *Defining criticality*

Knowledge mapping obtained during the previous stage defines the distribution of different areas of knowledge. We must now reveal the critical character of these areas of knowledge.

Area criticality is defined as an assessment of the risks/opportunities created by a particular area for the company. For example, there may be risks of losing knowledge or know-how, which may have damaging consequences. There may be grounds for developing the area to obtain benefits or advantages for the company (increased productivity or market share, etc.).

This therefore involves defining both what could be "objectively" the criticality of certain knowledge and an assessment method capable of identifying the most critical areas of knowledge in the mapping.

11.5.2. *The criticality study as risk assessment*

The criticality study can be viewed as an assessment of risks. In this respect, it is similar to the overall risk study referred to in section 11.1. In general terms, a risk is defined as an element or a situation that may prevent the goals of a company from being reached. A risk is usually defined in terms of gravity (consequences) and the probability that the situation will occur. Risk assessment can be expressed in terms of KM by referring to the criteria introduced in the previous section (Figure 11.6).

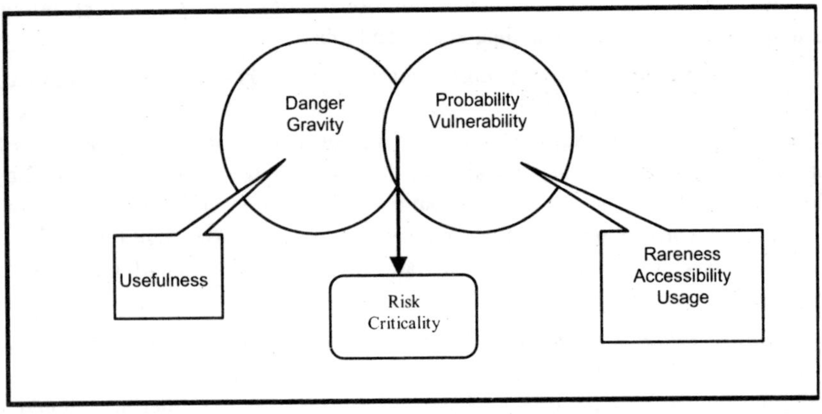

Figure 11.6. *Risk assessment*

11.5.3. *Defining critical factors*

Critical factors for determining the strategic importance of a corpus of knowledge are highly dependent on company culture and position. A collection of criteria, to which companies can refer when assessing criticality, has been created.

Critical knowledge is defined as *knowledge which is of value, but which is difficult for the company to exploit.*

To be of *value,* knowledge must first be *rare*. This may be the case when the company is the sole possessor of this knowledge (leadership in the field, originality and confidentiality of knowledge) or when this knowledge exists externally, but is inaccessible. This can also be due to the fact that there are few possessors of knowledge within the company and/or that these possessors are relatively unavailable. Finally, knowledge acquires value when it cannot be substituted, in other words when it is impossible to replace with other knowledge allowing similar results to be obtained.

However, if it is to be of *value,* knowledge must also be *useful*. It is useful because it concerns an area considered strategic, in other words an area appropriate to the business operations and end-purposes of the company and creating value for the stakeholders (customer or beneficiary satisfaction, stock-holder interest, employee interest, community interest). An area that is emerging and likely to generate value in the long term can also be considered strategic. Knowledge that can be reused in another area (area reusability and adaptability to needs) is also knowledge that can be useful to the company.

Frequently, a company unconsciously manages a large volume of knowledge that is at its disposal. Critical knowledge is therefore knowledge of value, but which *is not or is little exploited,* either because the company has not become aware of this asset or because this knowledge is difficult to exploit. The company must first be capable of harnessing its knowledge in order to exploit it. A first difficulty involves identifying sources of knowledge (people or documents). Access to knowledge may require creation or mobilization of contact networks. A second difficulty results from the fact that knowledge exists essentially in tacit form and evolves rapidly.

Optimum exploitation of knowledge presupposes not only a capacity for harnessing knowledge, but also a capacity for *integrating it into applications*, which is all the more difficult when knowledge is deep and/or complex, when it is difficult to appropriate and when area background knowledge is necessary in order to understand this knowledge. Finally, knowledge implementation often requires in-depth knowledge of the environment and a wide network of (internal and/or external) relationships.

Table 11.2 summarizes critical factors.

Thematic axes	Criteria
Rareness	• Number and availability of possessors • Specific (non-subsidiary) character • Leadership • Originality • Confidentiality
Usefulness to company	• Appropriateness to business operations • Creation of value for stakeholders • Emergence • Adaptability • Reusability
Difficulty in harnessing knowledge	• Difficulty in identifying sources • Mobilization of networks • Tacit character of knowledge • Importance of tangible sources of knowledge • Rapidity of evolution
Difficulty in exploiting knowledge	• Depth • Complexity • Difficulty of appropriation • Knowledge background • Environmental dependency • Internal relational networks • External relational networks

Table 11.2. *Critical factors*

11.5.4. *Preparing an assessment schedule and performing the criticality study*

Each criterion is intended for assessment and, for this, we set up an assessment scale for each one using the European Foundation for Quality Management assessment method [EFQM, 99].

Each criterion is assessed according to a scale comprising four levels, representing the performance rating for the criterion. Each criterion assessment is based on one question. Each level is expressed by a clear summarizing sentence avoiding terms that are vague and that lead to confusion ("rating description"). Scoring can therefore be established criterion by criterion.

Area criticality assessment involves allocating a score for each area based on each criterion. The more critical the area, the higher the score. Each area is assessed separately from the others.

Restoration of results is graphically summarized in a "radar" diagram (see Figure 11.7).

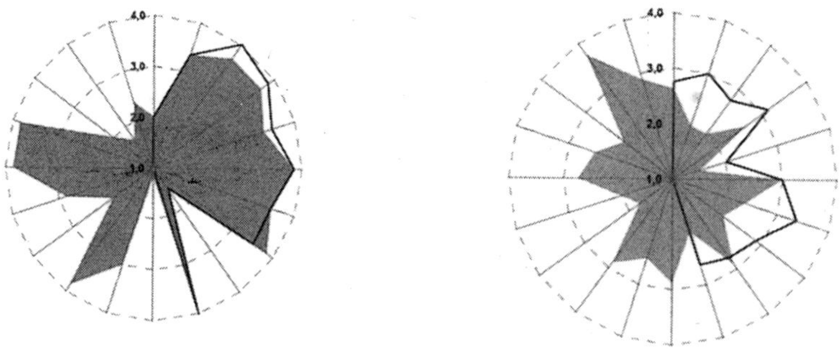

Figure 11.7. *Criticality diagram for two areas (at time t and at 10 years)*

11.5.5. *Calculating area criticality*

11.5.5.1. *Different criticality average values*

For each area, several agents can be involved in assessment. The assessment principle involves different assessors allocating a score to all criteria. Calculation of the assessment average value facilitates criticality analysis at different levels (overall average, per assessor, per assessment criterion). Calculation of an overall criticality for an area is done by averaging all assessments:

– *Area criticality per assessor based on one criterion.* Based on each criterion, the criticality for each assessor involves calculating the average assessment of indicators grouped into criteria.

– *Area average criticality based on one criterion.* Based on one criterion, the area criticality is calculated by averaging the scores allocated by each assessor.

– *Area average criticality.* Calculation of an area average criticality corresponds to the average criticality value for each criterion.

Calculating the different averages by changing the value of the weighting coefficient enables area classification and sorting to be performed according to several parameters (overall criticality, per criterion, per assessor, etc.). Other analysis data emerges from the table. Cross-linking of different data also facilitates detection of assessment biases.

11.5.5.2. *Assessment differences*

Comparison of each assessor's scores for the same area can reveal differences of viewpoint. These differences may be due to several factors, especially different assessor interpretation of criteria or different use of area knowledge depending on the department.

Assessment heterogeneity provides, in the first instance, the subject of discussions between the different people involved, in order to explain these viewpoints, formulate a common view of the area and even achieve a consensus on scoring.

Calculation of the average assessments of all areas for each assessor then allows any resulting assessment differences between assessors to be highlighted. Score heterogeneity reveals differences in interpretation as well as over- and under-assessments for areas of knowledge.

11.5.5.3. *Non-discriminating criterion*

A criterion or indicator can be considered to be non-discriminating if scoring variations in this parameter are low. In other words, if one (or several) assessor(s) give the same indicator score for a wide majority of areas, then this indicator is irrelevant to criticality assessment.

Criticality assessment is performed based on an assessment schedule. Selection of managers who are to be requested to fill in this schedule and data collection procedures may be a delicate issue. Efficiency and relevance must be simultaneously targeted, but overloading of line managers must be avoided. Communication operations and simplified analysis documents are very useful aids to the success of the operation.

Collected analysis data can form a considerable volume (for just one INRS research center, there were approximately 60 areas assessed by at least 3 managers using 25 criteria, representing some 4,000 collected data). A database and a spreadsheet program are therefore very useful for processing and representing these data, especially in a radar form (Kiviat diagram). A mapping representation tool so that different criteria can be viewed is essential for summarizing and presenting results.

11.5.6. *Analyzing results*

Overall or per criterion classification of areas of knowledge enables *priorities* to be derived. A *timescale* can be applied and areas of knowledge can be represented at both the assessment time and over a 10-year timescale. It is also interesting to *cross-link areas of knowledge* in pairs. Thus, cross-linking usefulness and rarity in 10 years' time provides a representation of a vision of the company over this timescale. This analysis operation leads ultimately to determining *growth areas, areas to be supported* in terms of difficulty in harnessing knowledge and *areas to be developed in partnership.*

11.6. How to manage a mapping project

Managing knowledge and know-how contributes to managing change within the company; the methods adopted must consider this dimension. Execution of knowledge mapping falls within the scope of this problem. Conventional project management aspects will not be considered here; rather, a number of specific points worthy of special attention will be highlighted.

11.6.1. *Opportunity study*

Mapping execution is highly dependent on the company context. Before deciding to run such a project, we should try to:

– situate the issue at the meeting point of company functions;

– understand the influential relationships;

– identify constraints and non-negotiable issues;

– identify sources of uncertainty;

– identify value-related tension relationships.

The project will then be formalized in terms of goals and priority actions.

11.6.2. *Managing change*

Knowledge mapping performance should be managed as a real project taking into account management of the change it causes. Management of change methods used in the company should therefore be mobilized. Specifically, this type of project, which requires heavy personal involvement, will be considered to require management of *change built up collectively* with those concerned, unlike management of *change prescribed* in terms of precise goals assigned from the start.

Resistances induced by possible consequences of the analysis results should not be underestimated.

11.6.3. *Action principles*

Collectively built-up change presupposes that the *values* and *action principles* underlying project management are explicitly and openly addressed right from project inception. These action principles must form the subject of a *debate*. For example, the following points should be examined: *ethical principles* involving the consequences of decisions made, procedures for *validating* collected information, *openness* in the methods of collecting, processing and making available information.

11.7. Conclusion

The different knowledge mapping-based methods and tools presented in this paper do not claim to provide an exhaustive final solution to the varied and complex problems represented by the process approach, by knowledge/competency management, by business framework implementation, by career management, etc. They attempt simply to show that this tool can facilitate producing coherence in these different approaches.

Perhaps one of the most interesting aspects of the approach is that mapping represents a strategic-level entry point to knowledge/competency management. It requires involvement of general management at the highest level of the company and represents a method and tool additional to those already existing in this field.

However, further research appears to be necessary. In particular, it would be worthwhile extending current work on the representation of collective knowledge associated with major organizational processes.

11.8. Bibliography

[Aubertin, 03] Aubertin G., Boughazala I., Ermine J.-L., Cartographie des connaissances critiques, *Revue des Sciences et Technologies de l'Information*, RIA-ECA series, col. 17 no. 1-2-3, pp. 495-502, Hermès Science, 2003.

[Authier, 92] Authier M., Lévy P., *Les arbres de connaissances*, Ed La découverte, 1992.

[Club, 00] Club Gestion des Connaissances, J.-L. Ermine, H. Laude: *A Knowledge Maturity Model*, Proceedings of "Knowledge Management; Theory and Practice" workshop (J.-L. Ermine ed.), in *PKDD2000 (Principles of Knowledge Discovery from Data)*, Lyon, pp. 13-1812, September 2000.

[ECC] *Club gestion des connaissances*, CD Rom edited by Club Gestion de la Connaissance, INT éditeur, 2004.

[EFQM, 99] *Le modèle EFQM d'excellence*, European Foundation for Quality Management. 1999.

[Geffroy, 02] Geffroy F., Tijou R., *Le management des compétences dans les entreprises européennes: les différentes approaches*, Insep Consulting editions, Paris 2002.

[Grundstein, 00] Grundstein M., "From capitalizing on company knowledge to Knowledge Management", in *Knowledge Management, Classic and Contemporary Works* by Daryl Morey, Mark Maybury, Bhavani Thuraisingham, The MIT Press, Cambridge, Massachusetts, pp. 261-287, 2000.

[Grundstein, 00] Grundstein M., Rosenthal-Sabroux C., "Capitalisation des connaissances de l'entreprise et aide à la decision", *Revue 2000 UE ENSAM*, Dunod, Paris, pp. 242-250, 2000.

[Jolis, 99] Jolis N., *La compétence au cœur du succès de votre entreprise*, Editions d'Organisation, 1999.

[Lichtenberger, 97] Lichtenberger Y., La compétence comme prise de responsabilité, Communication at ETMT seminar, Dijon, October 1997.

[Lorino, 01] Lorino P., *Méthodes et pratiques de la performance: le pilotage par les processus et les compétences*, Editions d'Organisation, Paris, 2001.

[Lorino, 92] Lorino P., *Le contrôle de gestion stratégique, la gestion par les activités*, Dunod entreprise, Paris 1991; new corrected edition March 1992.

[Penso-Latouche, 00] Penso-Latouche A., *Savoir-être: compétence ou illusion?*, Editions Liaisons, Paris 2000.

[Tarondeau, 98] Tarondeau J.C., *Le management des savoirs*, PUF, Paris, 1998.

Chapter 12

Knowledge Management and Innovation (Innovation Maturity Model)

12.1. Introduction

Knowledge Management (KM) consists of the organization of strategic corporate knowledge: its pinpointing, protection, valorization and creation [Dieng, 2000].

KM is a strategic approach for companies, structured within fundamental processes which are for the moment variously developed, according to the maturity of organizations in this area and the advancement of methodological and technological offers [Ermine, 2000a].

A fundamental expectation of KM lies in the field of innovation. In today's economic world, corporate survival relies upon constant and sustained innovation. Innovation represents henceforth a major competitive stake within corporations which accordingly seek to accelerate its underlying processes. In current acceptance, innovation covers two very different processes: innovation as the process leading from the "idea to the product"; and another process, more internal and higher up in the enterprise, which we will call "creativity". Mechanisms for accelerating innovation are of a very different nature according to the process considered. In the first case, structural and logistic supports ("innovation assistance") are of particular concern. In the second case, KM is one of the levers at the service of a voluntary policy of support to creativity, which conditions the consistent and efficient

---

Chapter written by Jean-Marie BEZARD.

evolution of the corporate culture of knowledge. It is this second point that concerns us here.

12.2. The evolution and path dependence hypothesis

The fundamental hypothesis set out here is the identification of the process of creativity within a process of evolution of the corporate knowledge capital. It is a hypothesis that may seem naïve at first (the creation of new knowledge enriches knowledge capital), but which in fact is in opposition to certain beliefs ("it's better to forget what one knows in order to be innovative") and numerous practices (no wasting time analyzing the past and/or the present to find new ideas), more closely linked to hypotheses of "spontaneous generation" than to hypotheses of evolution.

Another hypothesis proposed here is that innovation is a process of "endogenous and cumulative technological creation", which is to say that it is the very nature of the culture of knowledge accumulated within an organization that predetermines the path of evolution of this knowledge (and even of the organization itself). There is neither pure creation nor creation dictated solely by external constraints, but an evolution of ideas, through assimilation, accommodation, mutation, etc. This evolution of ideas within the corporation stems from its "genetic culture", which is constituted, among other things, of its culture of knowledge. It is therefore existing knowledge that conditions future ideas, and thereby leads to innovation. This hypothesis (known as "path dependence") [Coriat, 1997; David, 1994] demands therefore the analysis of the evolution of a system of knowledge relating to its history in order to better master its future evolution

From this point on, KM becomes a tool for innovation. Indeed, KM cannot be solely reduced to the transmission of information. The capitalization of knowledge and the analysis of critical corporate know-how permit the discovery of those laws of evolution specific to each enterprise, with the aim of bringing to light the combinations that prove to be the most innovative in each situation. This procedure also allows the enterprise to open up to new knowledge that could be effectively hybridized within the enterprise's environment.

A good understanding of the intellectual capital of the enterprise and its environment therefore favors the emergence of new ideas. A historical analysis of ideas can lead to the discovery of "laws of evolution" unique to the enterprise, providing guidelines for future evolution (discovery of innovation factors, unexplored paths, etc.). Work done on past organization reveals the establishment of conditions that trigger innovation; a good coordination of internal knowledge and know-how with environmental information (monitoring, economic intelligence, etc.) may favor creativity.

12.3. Innovation factors

A work commission based on these strong hypotheses was formed within the Club Gestion des Connaissances (Knowledge Management Club).

12.3.1. *Case studies*

The objective of this work group was to determine the key innovation factors within an organization, which may also be detection and piloting factors for this process. The method chosen to do this was based on observation and case studies. An analysis grid was defined from the start, comprising a certain number of points to report on:

– the history of the innovation (the purpose of the innovation, its course, the historical context, the players);

– the external environment (the monitoring process, general environmental constraints, external networks);

– the internal environment: internal culture (the body of references to the innovation studied, previous innovations in the enterprise, internal networks);

– the process of emergence and selection of ideas.

Six member organizations of the club then defined a product, service or process recognized as innovative, or as having been innovative, and identified an individual or group of individuals capable of telling the story of the innovation: the city of the children, the city of sciences and industry, the multifunction house at PSA, the Cortex project in Cofinoga, the ZX rear wheel-axle unit in PSA, the innovation in bonding with SMEs (small and medium-sized companies) at Thales and the DGA (the General Direction of the Armament), a laser technology at ONERA (the National Office of Studies and Aerospace Rcsearch) [Benhamou, 2001]. These cases were examined using the analytical grid.

12.3.2. *The Innovation Maturity Model (IMM)*

In order to construct an operational synthesis of this study, the technique of "Maturity Models", developed within the club [Club, 2000; Ermine, 2000b], was used.

A maturity model is composed of an ensemble of criteria, grouped in broad significant categories. Each criterion has a descriptive name, and an associated question allowing the understanding of its role within the given problem. The evaluation of each criterion is measured on a scale of 1 to 4, with a higher ranking corresponding to greater maturity of the organization in relation to the given problem. Each level is described in such a way as to facilitate evaluation, eventually including the giving of examples. In order to permit a consistent evaluation of the grid as a whole, a system of reference is established, providing possible axes of progression in the maturity of the organization (for example, from the individual to the collective, from the constrained to the spontaneous, from information to knowledge).

The purpose of a maturity model is to evaluate the ensemble of criteria and to synthesize the maturity of a given organization (enterprise, department, unit, etc.) in the domain considered.

This model, the IMM© (*Innovation Maturity Model*) [Club 03], consists of 21 criteria, grouped in four categories (thematic axes):

– policies for innovation (capability of inquiry, R&D policy, professional career management policy, etc.);

– organization and facilitation of innovation (search for heterogeneity of individuals within the enterprise, impact of the organizational structure of the enterprise on innovation, capability to generate degrees of freedom, etc.);

– attitudes favoring innovation (capability of assimilation, capability of accommodation, capability of managing breakthrough, etc.);

– cultural factors (constraint management, capability of calling into question existing situations, development of individual and collective attitudes favoring innovation, etc.).

Each criterion is rated on a scale of 0 to 3. To facilitate analysis and rating, brief rating descriptions for each criterion are used instead of normative descriptions.

The scale of 0 to 3 corresponds to company perception of the criterion according to "semantic axes", which provide the frame of reference for assessment:

– 1st axis	:	from Withdrawn	to	Open-minded
– 2nd axis	:	from Mistrust	to	Confidence
– 3rd axis	:	from Unorganized	to	Organized
– 4th axis	:	from Opportunism	to	Strategy

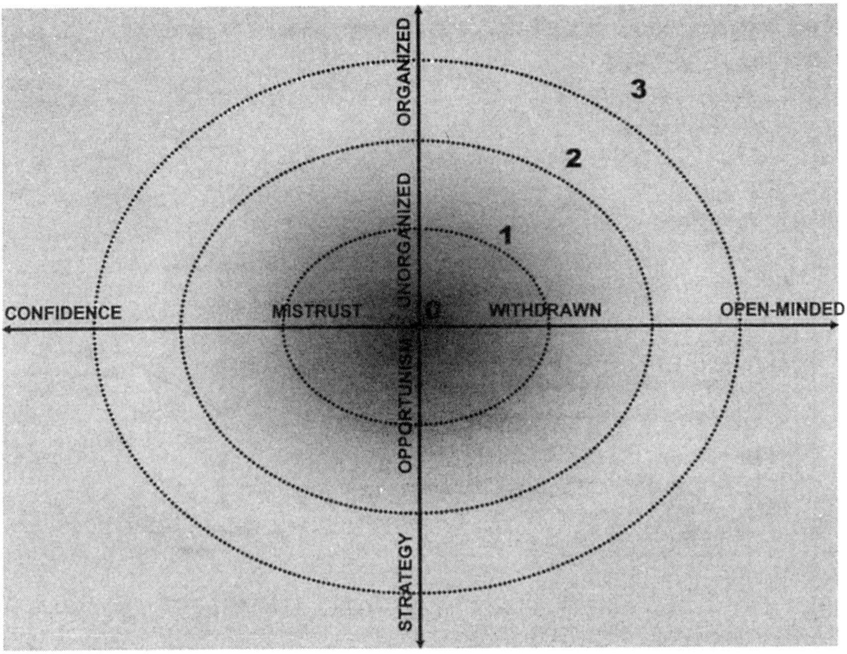

Figure 12.1. *Assessment framework for the criteria of the IMM*

For example, in the first class, a criterion and its levels of evaluation are as follows:

– *criterion:* capability of integrating external knowledge;

– *question:* how does the enterprise use information detected in its environment as being useful?;

– *the levels:*

- *level 0:* passive attitude – *general mistrust regarding information obtained*;

- *level 1:* use of information push – *information is diffused in a targeted or untargeted manner and without follow-up. Important difficulties in defining the use of the information obtained in operational, tactical or strategic terms*;

- *level 2:* organization of information qualification – *information obtained provides the opportunity to organize reactions, questioning, sharing, etc. Information is qualified and exploited;*

- *level 3:* use of information as an action lever within the company – *collective "sense making" is planned. Appropriate measures are envisaged and under discussion. Reporting is done at a strategic level.*

The IMM model provides, on the basis of information for all criteria, an image of the company, in its capacity to generate innovation. It is currently in the process of test and of validation.

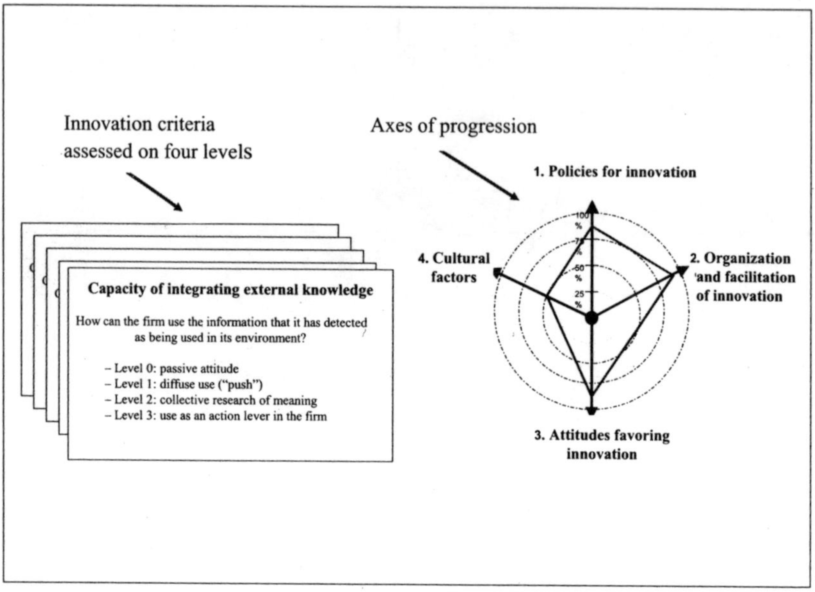

Figure 12.2. *Innovation Maturity Model*

In conclusion, the path dependence hypothesis has helped us to identify determining innovation factors in a coherent way, allowing us to trace and evaluate an "innovating profile" of the company.

12.4. Conclusion

Innovation is from now on a major competitive stake in companies that seek to control and accelerate the processes which underlie it.

Among the accelerating mechanisms of innovation, one that is explored very little is a thorough analysis of the past, which seems paradoxical! However, the hypothesis known as "path dependence" defends the thesis that new knowledge, and therefore innovations, are born from a process of cumulative evolution of the knowledge capital. Just as for the living systems, the evolutionary phenomena of change, accommodation, assimilation (which correspond to various types of innovation – rupture, incremental, etc.) start from this true genetic capital, which is the knowledge capital of the company. It is, however, important to be securely based on existing knowledge and its evolution in the past, in order to more effectively approach future evolution.

KM is another emerging strategic problem for companies. It recognizes a decisive value that is added to the knowledge capital accumulated by the companies. It seeks to develop this capital.

We have just seen that these two problems, when considered in terms of the formulated hypothesis, are far from being paradoxical, and that it is possible to combine them in a single, even step.

12.5. References

Benhamou P., Ermine J.-L., Taran J.-P., Tounkara T., Waeters A., Evolution des connaissances et innovation, application à une technologie laser à l'ONERA, *Extraction des connaissances et apprentissage*, no. 1-2, pp. 279-290, Hermès, 2001.

Club Gestion des Connaissances, Ermine J.-L., Laude H., A Knowledge Maturity Model, Proceedings of "Knowledge Management; Theory and Practice" Workshop (J.-L. Ermine ed.), in *PKDD2000 (Principles of Knowledge Discovery from Data)*, Lyon, pp. 13-18, 12 September 2000.

Club Gestion des connaissances. Innovation Maturity Model, June 2003.

Coriat B., Weinstein O., *Les nouvelles théories de l'entreprise*, Collection "Références", Livre de poche, 1997.

David P.A., Foray D., Dépendance du sentier et économie de l'innovation: un rapide tour d'horizon, *Revue d'économie industrielle*, no. exceptionnel, Economie Industrielle: développements récents, pp. 27-52, 1994.

Dieng R., Corby O., Giboin A., Golebiowska J., Matta N., Ribière M., *Méthodes et outils pour la gestion des connaissances* , Dunod, 2000.

Ermine J.-L., Les processus de la gestion des connaissances, *Extraction des connaissances et apprentissage*, no. 1-2, pp. 17-20, Hermès, 2000a.

Ermine J.-L., Le KMM, un premier outil pour aborder la gestion des connaissances dans une organisation, Knowledge Management, *Revue annuelle des élèves des Arts et Métiers*, pp. 120-125, 2000b.

Chapter 13

Technology and Knowledge Management (Technology Maturity Model)

13.1. Introduction

Information and Communication Technologies (ICT) are simultaneously the cause and the effect in the growth of knowledge. For this reason, they play an important role in Knowledge Management (KM) but do not constitute a panacea and their establishment must be accompanied by determining actions on the part of management. Unlike some choices of implementation solutions, the selection of KM tools cannot make do with a compartmentalized classification according to the functions of the software. It is more a question of relying on a global vision of the organization rather than focusing on software categories that are not always stripped of marketing effects. This vision must be connected to an infrared image, pointing to knowledge flows as well as to obstacles to their diffusion, from which optimization, elimination and circumvention solutions may be built. More than for other more classical domains, these solutions must be based on organizational, human and sometimes technological actions, built with coherence and with attention to the collaborators' motivation. The "Technology Maturity Model" of the Knowledge Management Club aims to provide basic elements for elaborating this organization landscape.

Chapter written by Olivier LEPRETRE.

13.2. Knowledge diffusion vehicle

Lisbon, March 2000: at the end of its Council meeting, Europe announces that it wants to be "the most competitive and dynamic economy in the world".

Mainz, 1448: absorbed in the contemplation of the freshly inked page coming from his new press, Gutenberg undoubtedly does not imagine that he has just started the chain reaction that will determine the European objective of the 21st century.

The path that links these two dates is that of knowledge. Since the appearance of language, about 100 million years ago, knowledge has been transmitted orally, or much later (5,000 BC), at the copy speed of scribes and monks. And, as for a nuclear reaction, printing will create the critical mass necessary for the beginning of the process. Each piece of knowledge could be communicated quicker and further, to be used or improved by other men, inspire them with new knowledge, just as each neutron gives birth to several others in the nuclear explosion reaction chain.

The history of inventions and discoveries, revealing the knowledge level of a civilization, shows us that their growth follows an exponential curve, which rises little after the appearance of the printed book, at one time so nicely called "the Renaissance". A reader of the very first print, Pic de la Mirandole, arrived at the only moment when he could choose "knowing all that is known" as a motto without too much exaggeration, because of the confinement of knowledge, which had limited its volume, as a result of lack of support. Two centuries later, the Century of Light started in Europe, the first tangible sign of the changing paradigm.

If so much knowledge exists today, it is not because we suddenly became cleverer than our ancestors. The human brain has not changed for several millennia, but for nearly six centuries its power has been built upon the acquisitions of preceding generations. It can then add its own contribution and transmit the result to its successors. In recent years, the knowledge diffusion cycle has been shortened to the extreme because any thought can incorporate contemporary pieces of knowledge wherever they are located in the world and even only a few seconds after their publication[1].

Undoubtedly, the most famous scientists would not have made the same discoveries if they had been born earlier. Could Albert Einstein have developed relativity theories without Maxwell's equations and Lorentz's transformations? Would Louis Pasteur and Robert Koch have made the same discoveries without the preceding work on bacteria, not to mention all the elementary knowledge they acquired from their youth? Conversely, what would the mathematical genius of

1 Google Alert: http://www.google.com/alerts

Srinivasa Ramanujan have found if his theories have been based on peers' knowledge instead of spending part of his short life reinventing the necessary theories? The history of inventions shows us how knowledge feeds knowledge. After the machine helped the creation of "ready to wear", numerical technologies are forcing us to consider the "ready to think". Each person can thus directly use increasingly focused concepts by handling complex knowledge, representing some dozens of years of work and experience, in the form of elementary blocks accepted once and for all, or at least until the next discovery. And KM is here to help him or her.

After printing had been invented, improvements and discoveries linked together in the domain of communication technology. The tape recorder, the video tape recorder, cinema, telephone, radio and television not only amplified knowledge transmission but also, and implicitly, its memorization. Thanks to them, the computer and the Internet appeared and now absorb and bring together all those techniques in a numerical whole.

Knowledge can be available simultaneously anywhere on Earth in the form of speeches, documents, pages and messages. It enables task forces to exchange their ideas, researchers to share their results, isolated individuals to work with an intellectual network, students to question their teachers, consumers to know about markets, etc. Thus, as an example, vaccines may sometimes be developed within a few months, thanks to the collaboration of laboratories scattered over the planet in a neuronal network, whereas it would take years for a sole researcher.

Being both the cause and result of the knowledge explosion and the landscape change that they generate, numerical technologies transcend printing and are the allies of knowledge. They are instantaneous vehicles of transmission, memorize much (but not all) information more powerfully than the brain, add multiple dimensions to consultations, treatments and presentations. Transmitted orally for a long time, today knowledge is transmitted more and more by means of a numerical support.

When, in 1976, Herbert Simon proposed the principle of "bounded rationality", he highlighted an essential operating mode of the human being, which search engines and more generally KM will impact more and more deeply. The Nobel Prize winner explained why, when confronted with a problem, man does not choose *the* best solution but *his* best solution. Limited by time, comprehension capacities and accessible information, as well as moral and more subjective thoughts, we stop our research at the conclusion that appears to be the most satisfactory to our eyes. In other words, if it were possible to consult more relevant information in a similar period of time, there are strong chances that a more relevant solution would be identified or created. That is to say that technologies not only enable us to know more but they could enable us to know better.

Moreover, computers do not only store knowledge to put at our disposal but are also able to process part of it in a dynamic way, in the form of rules and data-processing codes. While capable of replacing the brain in a number of tasks such as calculation, drawing and production, computers show snatches of reasoning. It does not matter that a machine plays chess, extracts a square root, traces a circle with two points in a limited and repetitive way; it just does it. Only the result counts and a lot of knowledge is held today by machines which fulfill their role better than the majority of humans, for example, welding and painting automobile robots or planes that land without visibility.

When Ariane or Endeavour lifts off, the knowledge is no longer in the brains of men but inside computer memories. In everyday life, computer antivirus software contains a lot of knowledge that enables it to detect the "signatures" of those harmful programs that no one could memorize. For this reason, after having carried too many premature hopes, expert systems are returning in force to support simple but too huge a knowledge for the human spirit. A cookery recipe is not formulated in such a different way from a programming language and it would probably be possible for computers to understand such simplified language before long. It thus appears necessary not to consider solely tacit and explicit knowledge but also coded knowledge whose volume and place in everyday life is increasing.

Through computers, we have the very first unconstrained way of representing knowledge. It is evident that knowledge is dynamic while its representation, supported by numeric or paper pages, is rather static. The photograph is not the landscape, but only its representation from a particular point of view defined by a place, a distance, a moment. As soon as a parameter changes, the photograph is no longer the same and, in the same way, knowledge changes according to time and needs. The power of the brain is to be able to manage dynamic knowledge, adapting zoom and angle to the need, that is to say, the number and the granularity of required concepts. The brain contains both engine and information, intelligence and tacit–explicit knowledge. This explains the success of search engines and hyperlinks, which are the first step to a dynamic view of knowledge, suited to each need, by a system other than the brain.

Numerical technologies extend the capacity of the spirit capacities of human beings, just as machines extended their muscular force in past centuries. What is true for humanity is also true for an organization, which, through its computers, networks, software and the Internet, has billions and billions of neurons potentially connected in a huge knowledge system, free of space and time. But is this potential really exploited? Of course not, and it is the KM challenge.

13.3. The limits of the diffusion of knowledge

Such impacts encouraged the implementation of communication technology inside companies and this step was obviously taken a long time ago. Is it necessary to continue, following technical evolution? Is this the panacea for KM?

Obviously prudence is essential, for several reasons: first, because technology is far from being omnipotent and has consequent limits; second, because a number of obstacles are related only to human beings; and finally because, when dealing with more and more knowledge, technology has a deeper and deeper impact on the daily life of individuals and their relationship to work. Replacing muscular force was not, and still is not, done without impact and, even if the replacement of the brain is far from being scheduled, intervening in its domain without taking precautions is a bit like opening a Pandora's Box; one cannot always predict what will happen. The minimum will be to think carefully about the roles that are shared between the worker's brain and its electronic substitute.

First of all, let us speak about the limits of technology. There is no learner technology, because memorizing does not mean learning. One can memorize a whole text without being able to draw anything, as a computer does. Learning means modifying one's mental schemas, one's ways of thinking, so as to act differently thanks to experience feedback, thanks to new knowledge. Of course, some software is able to provide different representations according to the user; neuron networks have some training capacities, but this is only the beginning. Unlike humans, information technology is still unable to adapt to a new context, unless it is strictly planned, and this is its major limitation.

Context, shape and semantic recognition are also out of the reach of information technology. There is, of course, software that analyses sound and video, and data-mining tools whose capacity to highlight certain phenomena as the sources of an epidemic be compared to a creation of knowledge, but they are not very widespread. This limit notably burdens the exploitation of the huge storage capacities by not allowing an optimized search that takes into account the meaning of words and sentences, the contents of images and speeches ... Because of these limitations, one can say that "too much information kills information" and the principle of "bounded rationality" applies. While working only with words and not on meaning, search engines swamp relevant answers with superfluous ones, often copied one from another, more or less far removed from the problem, often even irrelevant. It is because of this semantic incomprehension that automatic translation software sometimes gives results that are amusing but completely unusable.

Though huge volumes have been capitalized, technology does not help much in the capitalization of knowledge itself. Brain capacities can be limited in terms of rough memorization, that of a directory, for example, but they are extraordinarily good at relative memorizing, that is to say contexts, situations and experiences. The difficulty in transmitting information to a computer is thus a limit not only of its functions but also of the human spirit, which needs much structuring effort to capitalize on this. This reminds us Michael Polanyi's famous phrase, "We know more than we can tell" [POL 66]

We should add to software's disadvantages that those said to be designed for "knowledge management" are legion. If they were fewer, that would hardly change the problem, as no software can really been excluded from this concept, because the majority of the programs can claim to play a role. A word-processor and email software help to capitalize, CAD software assists the user in his choices more and more, etc. There are hundreds of products, the least expensive of which are not inevitably the worst, many of which will no longer exist in a few years, because they are too numerous. This established fact forces us to consider a majority of them with prudence because once integrated in given software, knowledge may appear difficult to transfer to another system. That is why a manager should not *think about buying software but about federating the software of the organization* (and perhaps buy one program) in order to the manage knowledge.

Despite these technical obstacles, which will doubtless evolve, it is indeed fundamental to consider the obstacles created by human factors. By definition, any kind of innovation upsets our habits, relationships and everyday life to some degree, even if the modification takes time and proves to be insidious, starting by prolonging an old habit. Thus, television was initially a culture-orientated filmed radio, before becoming a financial vector through publicity, SMS and derived objects, while the mobile phone prolonged the ordinary telephone before becoming the multi-function pocket knife allowing messages, games, photos, videos, localization, authentication, numerical payment, etc. and the computer was initially a typewriter before becoming what we now know, far less than what it will be tomorrow. Furthermore, that is why, in the KM domain, the computer is still used as a continuation of the book whereas its capacities go much further.

In our everyday lives, we make technological choices because we think that they will bring new pleasure or advantage. We are not obliged to yield to all temptation. The mobile phone spread because it constitutes a true evolution, but not everyone has one. The personal assistant is still not widespread because its effectiveness remains to be proven for some and so much depends on each individual.

Much more than an individual, an organization should not yield to the technological mermaids of marketing. In choice of a KM solution, the managers' role is complex because they must take into account not only software function but also the multitude of individualities of their collaborators, with their personal constraints and objectives, ways of thinking, varying degrees of culture, but whose adhesion and coherence of operating modes are essential to the organization's survival. These differences mean that technical solutions that were good for one organization will not automatically be good for others, because even if the product is the same, the obstacles are different. For KM, the real human obstacles must be completely included in the specifications.

Knowledge generates knowledge, like neutrons in a nuclear reaction. This is a fact, but in such an engine, the reaction is also moderated, controlled by graphite rods, which prevent racing and explosion. These graphite rods also exist in an organization but they play their role so well that its heart is sometimes closer to stopping than racing. Their control is much more difficult and their appearance rather like spontaneous generation.

Relationships of power and influence, disputes between collaborators or departments, geographical dispersion over sites, floors and offices, secret worship, computer networks with low output; dozens and dozens of organizational, human and sometimes technological obstacles slow down the diffusion of knowledge. The differences in languages themselves, whether real like French and English or artificial like trade vocabularies, are often the first obstacles. Moreover, it is striking that an episode of the Bible evokes this power of knowledge. In the Tower of Babel, God prevents men from fulfilling all their goals, particularly reaching the sky, by inculcating different languages in them.

In a large number of organizations, coincidence and the coffee machine are the only vectors of knowledge dissemination. Technology can cancel space and time, a simple partition or a sideways look can create insurmountable borders. Man himself becomes the principal obstacle. The collaborators in most organizations put considerable energy into reinventing what was carried out a little earlier or a little further but hindered by ignorance or low esteem. However, when an organization allows the exchanges between its members to decrease, it must realize that its added value gradually becomes the same as if someone was working alone on a project.

To solve the problem, managers must consider the "principle of the smallest action" and applied it to knowledge. They must then integrate the existence of dozens of obstacles to knowledge dissemination in their company and find a way to reduce the energy required to get round them. Some obstacles like geographical dispersion and volume of knowledge be got round with technology. Some have nothing to do with this technique but it can help to solve them, just as search engines help to establish bonds between collaborators involved in similar fields in successive periods, because incentives can be created. For others, like the lack of common values, disputes and conflicts, it is not in technology that one will find the solution. A solution that was good for a conflict-free company will be useless in a sister company involved in major disputes. That is why, understanding this, more and more software editors, anxious that their product should not be devalued by a third party, do not present it as a panacea but as an excellent solution, working at its best when it accompanies other approaches.

More than for any other technology, tools intended to assist man in his most gratifying and deepest properties, that is to say his know-how, must be the object of increased attention. The risk of a bad choice can be a loss of investment, a rejection and even, in the worst case, tears or strikes. Without going to such extremes, KM through software installation must be planned in terms of what else is going on and not be based on situations as they arise. It is this vision that must pilot the setting up of products in a company.

Unfortunately, the most widespread approach to software confines us to a vision of monolithic products, compartmentalized into categories. Terms such as word-processing, spreadsheet and email, which were common sense, are now to be replaced by groupware, collaborative work, localization of expertise, e-learning, documentation management, content management, workflow etc., categories in which software is sometimes different, but more from a marketing than a technological point of view. Although equipped with import/export or cut-and-paste functions, many products work mainly in an isolated, autocratic way, creating new obstacles to the exchange of knowledge.

We are beginning to be optimistic about software functions of various origins all working on the same data structures, such as XML, and so dream of replacing presently owned tools with a numerical office composed “à la carte” of components from different sources. This would make it possible to have some word processing, some collaborative functions, a few presentation functions and so on, without having to be concerned with juggling different software that can collaborate only with itself.

In the same way, the keys to an organization are distributed in various hands, which is logical but causes some problems when transverse knowledge has to be handled. Some of the knowledge required by commercial people is held in the engineering and design department; product design must benefit from the maintenance men’s feedback; well understood purchasing must help to determine the conception of products according to suppliers’ tenders, etc. It is imperative to have the greatest collaboration between all departments and this is the beginning and end of KM.

It is true that the capacity to exploit knowledge structures through various cognitive functions of reception, classification, memorization, reasoning and expression is the privilege of a single brain. It should be recognized, moreover, that the capacity for dialogue between men is sometimes worse than that of software, especially when they are grouped in a little structure called a company.

It is possible, however, to progress to establishing what has been called a learner organization, if it is considered that the true problem is not to say that a company requires tools for "collaborative work" or "localization of expertise" but to paint a broad picture of knowledge flows, their origin, their destination, their criticality, their vectors or obstacles and thus to improve their management by means of organizational actions, human and sometimes, but not always, technological. And this is the real goal for managers, not the one described by editors who are unable to consider each organization's specificities.

13.4. The need for global vision

We therefore conclude that, undoubtedly, software plays a major role in KM but that also, undoubtedly, it does not solve all problems as obstacles having nothing to do with the technicalities must be taken into account.

Of course, we already know that things will not remain the same and that numerical technologies will still advance, in particular to overcome their weak points. The personal computer is only 20 years old and the Internet in reality is not even that. Eighty years of creations and improvements were made between the first flight of Clement Ader and the first of the 747. Despite their sophistication, today we are probably dealing with the biplane of computers and the 747 or A380 of data processing is still to come. However, technology will never solve emotional problems and could rather create some. For an organization, it is a question of preparing for a better technology that integrates human and organizational actions so that their evolution will not create new upheavals but rather facilitate KM in a homogeneous whole including them and their collaborators.

With this potential and these limits, one can certainly detect a complementarity of the strong and weak points of the brain and technology. Stated in terms of cognitive functions, technologies are gifted in expression, treatment and memorization, but weak or non-existent at reception, training and reasoning, which are the fields in which the brain excels. Cognitive or intellectual functions can be divided into four classes:

– receptive functions allowing acquisition, treatment, classification and integration of information;

– memory and training allowing storage and recall of information;

– thought or reasoning concerning mental organization and reorganization of information;

– expressive functions allowing communication or action [BER 91].

The objective would be thus to regard the organization as only one single system of knowledge, combining in a coherent and optimized way the individual capacities of the collaborators through the capacities of software and information gathered in knowledge bases. Technology can then be considered as a cement contributing to the neuronal network of the company, just as bricks need cement to become a wall.

To be able to integrate simultaneously the product functions and the knowledge they support, the obstacles that they help to eliminate or circumvent and the evolution of the organization to facilitate their integration, it is necessary to change the approach, the perspective, the point of view. Knowledge should be visualized as it would be by an infra-red camera highlighting sources, flux, potentials and obstacles, integrating all the software acquired by an organization, irrespective of their category of origin but associated with the competencies they support.

13.5. The Technology Maturity Model (TMM)

In line with these ideas, the "Role of Technologies" commission of the Knowledge Management Club decided to try a new approach. This one should be less oriented to product comparisons and much more oriented towards the evaluation of the software contribution to the total circulation of knowledge by working out a "Technology Maturity Model" (TMM) [CLUB 04].

The aim is to consider the whole or part of the organization so as to show how its members can more or less act as a "super-expert" thanks to a coherent combination between software and collaborators.

The basic idea is to be more centered on the cognitive functions of human beings than on those of the tools, to be "knowledge driven" and not "software driven". To achieve this, TMM is based on a derivation of human being's cognitive functions, applied independently of the various software, to precisely highlight flows or partitions between them.

The model must help to answer how and which software combination allows the organization:

– to acquire and structure new knowledge;

– to memorize and find knowledge;

– to combine and organize knowledge;

– to push knowledge to collaborators at the right time.

It is a question, for example, of highlighting:

– how some specificities of electric wiring in the streets of a town, carried out by the engineering and design department with a CAD tool and mainly known by terrain workers, can be found and given to a new commercial organization carrying out an estimate on the sector;

– how very complex knowledge about performances and limits of a numerical geographic information system can be formalized to guide a non-professional user in the choice of the information required by this system;

– how knowledge expressed via text processing software can help to connect two collaborators, for example by the means of expressed needs and a document's author's name;

– how the contributions and flow from only one product installed on the majority of the company's computers, such as production management software, helps the organization in the circulation of its knowledge.

13.6. Following a TMM approach

To facilitate the decomposition, the human cognitive functions were transformed into functions more suited to software:

– reception and treatment functions;

– memorization and research functions;

– combination and organization functions;

– expression functions.

As these basic functions were still too generics, they were broken down according to unitary standard functions as showed in Table 13.1 at the end of this chapter, which constitutes an example of the TMM approach. These standard functions must then be analyzed in terms of the organization's trade functions.

Experience shows that it does not seem judicious to seek at all costs to preserve the original cognitive functions. In many cases, the brain seems to carry out at the same time the tasks of perception, reasoning and action. Even if a temporal decomposition is certain, it is not always relevant as regards the analysis and complicates the work. The objective is to reflect the global flux of knowledge and not to quote each software function, which is the classic way of tool comparison.

After, or in parallel with, this development, it is necessary to complete each function with a series of attributes, some being optional:

– software concerned;

– knowledge handled and its criticality;

– handling competences;

– departments, units concerned;

– context of use;

– potential or proven obstacles.

TMM can help the description of various potential improvements not just related to technologies:

– strong processing and search capacities but low capacity for creation and/or update;

– important capitalization but low or zero exploitation;

– absence of a connection between similar knowledge managed by different software;

– bottlenecks or partitioning between departments;

– identical functions provided by different software;

– missing functions to manage knowledge with a high potential;

– software that is under or over exploited;

– knowledge that is not managed;

– knowledge that is managed in a redundant way, structured in several different manners;

– badly managed knowledge (safeguards, confidentiality, etc.);

– knowledge that is too complex to be managed by software (thus potentially critical);

– needs for acquisition, development or replacement software;

– obstacles to diffusion;

– etc.

Bringing the TMM closer, knowledge mapping [AUB 03] will make the following possible:

– If TMM is elaborated first, to be used as an entry point to the mapping, identified knowledge can be supplemented by extending the perimeter to the knowledge not managed by software.

– If mapping is carried out first, to constitute an entry point by adding software to the identified knowledge or to identify unmanaged knowledge, and thus conclude:

- that a possibility for improvement is possible by the introduction of adapted tools;

- that knowledge is too complex to profit from software support, which can mean it should be included within critical knowledge (which in theory would have been identified in the mapping).

13.7. Application of TMM

Application of this approach is done in the following stages:

1. Define the perimeter, i.e. the departments.
2. Identify the organization's software.
3. Identify the unitary functions.
4. Identify the standard function with a trade function. For example, replace "to seek", with "to seek the isolating valves", "to supervise" with "to supervise the evolution of the standards", "to correlate" with "to correlate the expenditure between several sites" and "to recognize" with "to recognize a profile of the problem".
5. Associate the attributes.
6. Analyze the results and define actions.

The experiment demonstrated the following:

– The cognitive function is not inevitably "exclusively software" but often one of the pairing user/software. The goal is to highlight how technologies *assist* the collaborators and not how they replace them.

– It can be necessary to define several unitary functions to describe a process. It is necessary for example "to detect the modification of a Web site" and "to notify an evolution" to cover the phase of perception and the phase of expression. It is also possible to implicitly cover those two phases in only one function if their decomposition is not relevant (if, for example, detection is not memorized at a later stage). A decomposition is interesting if it makes it possible, for example, to highlight an obstacle between the two functions.

– Only the functions that relate to exchanges between software and/or collaborators are to be identified. Purely internal functions with a product are of no interest.

– The list can looks like some profiles of competences. This result is a good thing and is desirable since it facilitates convergence between various approaches in the knowledge analysis. However, it should in theory contain only the functions for which there is a software either on the market, in the organization or to be developed.

– The essential difficulty is in the granularity of the detail of the functions. It is absolutely necessary that these should be neither macroscopic, to be representative of the principal software functions, nor microscopic, to be drowned in a profusion of functions.

Table 13.1 is an example of some functions that would need to be associated with attributes.

Standard function	Function trade
To raise	To raise a point of measurement of air quality To raise a spelling mistake To raise a methodology error
To consign	To consign an experience feedback in laboratory To consign a customer contact
To seek	To seek the isolating valves To seek a chemical formula, a patent To seek an experience feedback, an example of a solution
To interpret	To interpret a satellite image To interpret a result of medical imagery To interpret a graph of measurement in microgravimetry
To supervise	To supervise the industrial risks legislation To supervise official announcements To supervise websites of competitors
To correlate	To correlate the expenditure between building sites To correlate epidemiologic measurement results
To recognize	To recognize the profile of a problem To recognize a typical situation To recognize the presence of several symptoms
To locate	To locate an expert in water treatment To locate a document
To detect	To detect the raising of a pump dysfunction To detect a programming error To detect a virus

To associate	To associate a need with an expert To associate similar documents To associate ideas
To translate	To translate French into English To translate observations to a graph
To qualify	To qualify a geography numerical product To qualify a result as a risk factor
To suggest	To suggest modeling of a calculation To suggest a standard component To suggest a standard answer
To hold or create a history	To create a history of successive equipment modifications To create a history of indicators
To combine	To combine space and temporal information to prevent an avalanche risk To combine the availabilities for establishing planning To combine newsletters
To filter	To filter emails, news, addresses of sites To filter documents To filter results of measurement
To analyze	To analyze an assessment
To select	To select a data file
To justify	To justify a result
To supervise	To supervise a process
To communicate	To communicate a customer contact To communicate a solution
To calculate	To calculate the path of an automobile electrical loom

Table 13.1. *Some functions that would need to be associated with attributes*

13.8. References

[AUB 03] Aubertin G., Boughzala I., Ermine J.-L., Cartographie des connaissances critiques, *Revue des Sciences et Technologies de l'Information*, RIA-ECA series, vol. 17, no. 1-2-3, pp. 495-502, Hermès Science, 2003.

[BER 91] Bérubé L., Terminology of neuropsychology and neurology of the behaviour, La Cheneliere, Montreal, 1991.

[CLUB 04] Club Gestion des connaissances, Technology Maturity Model, January 2004.

[DEL 03] Delahaye J.-P., Lille University – *Scientific American/Pour La Science,* no. 314/no. 12, 2003.

[POL 66] Polanyi M., *The Tacit Dimension*, Routledge & Kegan Paul, 1966.

PART 4

Case Studies

Chapter 14

Once-upon-a-time Knowledge Management at Mann+Hummel Automotive France

14.1. The increasing importance of knowledge sharing for Mann+Hummel France

Mann+Hummel Automotive France (MHFR) employs 400 people. Activities include administration and R&D (in Laval) and production plants (Laval and Grenay). MHFR has 50 years' experience in automotive plastic part design and production. At the beginning, the company expertise was mainly in plastics processing. Nowadays, MHFR has become a tier-one system supplier. One of its priorities is to preserve and manage its knowledge and to develop new expertise.

The Mann+Hummel Group is an international company. In 2003, the company achieved sales of approximately €1,140 million. The Mann+Hummel Group currently employs around 9,000 people at 41 locations worldwide. The company develops, produces and sells technically complex automotive components such as air filter systems, intake manifold systems, liquid filter systems and cabin filters for the automotive industry, and filter elements for vehicle servicing and repair. For general engineering, process engineering and industrial manufacturing sectors, the company's product range includes industrial filters, filter systems, and materials handling systems and equipment. Mann+Hummel's customers come from a large number of sectors, with series production for the automotive industry occupying a key position.

Chapter written by Nathalie LE BRIS.

In 2002, previously part of the Solvay group, the entity was purchased by Mann+Hummel group (M+H). This purchase required major exchanges with the other entities of the group: units with different organizations and languages. It also meant the introduction of new international customers and the beginning of international projects requiring more co-operative work.

The confidence of the automotive manufacturers in this innovating company enabled it to gain many markets. The number of plastic parts under the car bonnet is increasing and they are becoming more technical, which requires a good control of the technical risks. In parallel, total quality systems at customers require the avoidance of repeating mistakes.

As for any automotive project, new product development is carried out by simultaneous engineering. All project team members must have access to information very quickly. The project approach requires to effective working in a transverse organization. The concept of co-operative work takes on its full meaning in simultaneous project engineering.

At Laval, resignations of personnel are few and employee turnover is low. A sharing culture is well established. Thanks to its size and its history, direct transfer of knowledge (directly from a person to another one) at MHFR is carried out naturally. However, indirect transfer (using the intermediary of a document, a video, etc.) is harder; one can feel the lack of explicitness and inexperience in document sharing.

In 2002, Mann+Hummel Automotive France headquarters decided to launch a Knowledge Management (KM) adventure. They started by hiring a full-time knowledge manager for a better understanding of KM and to carry out the best KM actions for the plant. The aim of the directors was to avoid making the same mistakes again (without slowing down innovation) and to succeed in integrating the site into the group.

14.2. An approach based on core knowledge cartography

14.2.1. *What are the needs of the employees?*

MHFR activities are split into seven processes. The process called "new product development" includes much knowledge in movement. New knowledge is created every day because innovation is really strong in this process. As they needed to select a restricted sample to test KM actions, the directors chose this process representing 100 people from four different departments.

A survey through KMM grid (Knowledge Maturity Model from the Knowledge Management Club) was carried out on 32 people in order to collect their feelings about the strong and weak points of the company for KM. Five elements appeared through the analysis of their answers:

– a strong sharing culture;

– a lack of methods to formalize their experience;

– a lack of tools for sharing it;

– an organization favorable to KM (recruitment of a knowledge manager);

– a good moment for starting a KM approach.

This kind of investigation is really important for a KM project because the final aim of KM actions is to meet the user's needs. If they consider that they have enough tools, there is no need to buy other tools. Compared to many companies where the lack is in organization or culture, MHFR needs to offer employees means for sharing and explicitness: tools, methods.

14.2.2. *A cartography that allows us to build an action plan*

Mann+Hummel France wants to advance in KM, but what knowledge is available? Which knowledge is strategic for the company? How is it currently controlled? In a nutshell, which knowledge is it necessary to start with and how? In a company that has designed and developed technically advanced automotive products for almost 50 years, knowledge is various and is distributed to all the employees. It is absolutely necessary to start by clarifying the knowledge fields of a process in order to be able to build a capitalization and sharing action plan.

MHFR chose to use the very complete work of the Knowledge Management Club on the core knowledge cartographies with the help of CEGOS and its very fast and operational K3M approach (see Figure 14.1). A few days were enough to carry out knowledge cartography for these 100 people.

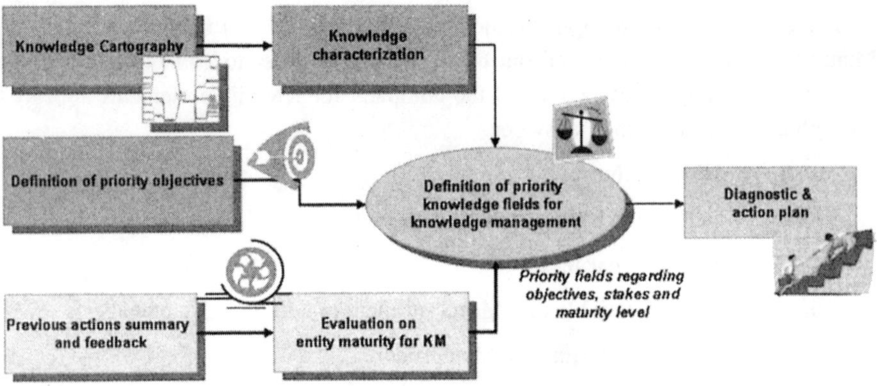

Figure 14.1. *CEGOS cartography approach*

The first step for a good cartography on fields consists of listing in a tree structure all fields of knowledge for a process. This list was drawn up on the basis of the interviews with 10 people (department managers and some other people with a clear vision of available knowledge) and by the study of various documents describing the company and the process. Knowledge fields were gathered according to four large branches or points of view (see Figure 14.2):

– knowledge about activities in this process;

– the tools/resources necessary to carry out activities;

– knowledge related to the final products of this process. For MHFR, we are talking about knowledge about parts designed by the company (manifolds, filters, cam covers, air ducts, etc.) and the production methods set up for these parts (welding tools, injection machine, etc.);

– transverse knowledge that can appear on several preceding branches (e.g. acoustics, thermodynamics, etc.). One of the main rules is that a knowledge field is only situated in one leaf of the knowledge tree. When a field is likely to be duplicated, it is added in one of the "transversal knowledge" branches.

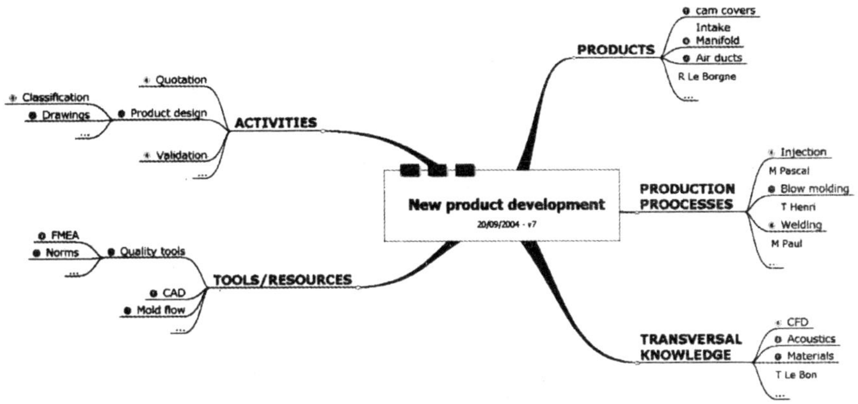

Figure 14.2. *Example of a knowledge tree. The MHFR one is much bigger and also confidential*

Once this list had been completed, we needed to give priority to those knowledge fields in agreement with the strategy of the company. In order to identify priority knowledge fields, it is first necessary to define the criteria that are the nearest to strategic axes of the company. For MHFR, four binary criteria were defined:

– Is this knowledge field frequently used?

– Does a better control of this knowledge field have technical influence on our products (innovation)?

– Does a better control of this knowledge field have an economic influence?

– Will a better control of this knowledge give us a competitive advantage regarding our customers?

All the knowledge fields were reviewed in the light of these criteria with the help of the technical director and the middle management. Each knowledge field was thus given a priority index (one of five indexes).

Then we wanted to know how these knowledge fields were controlled and how they must be controlled in the future in order to select a good KM action for each one. For knowledge priorities 1 and 2, we studied their current level of control, which corresponds to a combination of the following two parameters:

– The level of explicitness of this knowledge: is it formalized (explicit knowledge) or is it only in the experts' brains (tacit knowledge)?

– The number of owners of this knowledge: limited to some holders (rather individual knowledge) or widely dispersed (rather collective knowledge).

These parameters helped us to determine the current level of control shown by the four textures in Figure 14.3:

– individual tacit (squares): the worst level of control;

– collective tacit (horizontal lines);

– individual explicit (diagonal lines);

– collective explicit (with bubbles): the best level of control for a company.

Priority levels, as well as current levels of control, are represented on the knowledge tree by four colors (shadings) and five numbers in order to have an overall vision of available knowledge on only one tree.

The current level of control, as well as the target level, enabled us to create an action plan in agreement with the strategy of the company and in phase with the wishes of users. KM actions will not be the same for individual explicit knowledge (effort on document sharing for example) as for collective tacit knowledge (effort on explicitness, for instance).

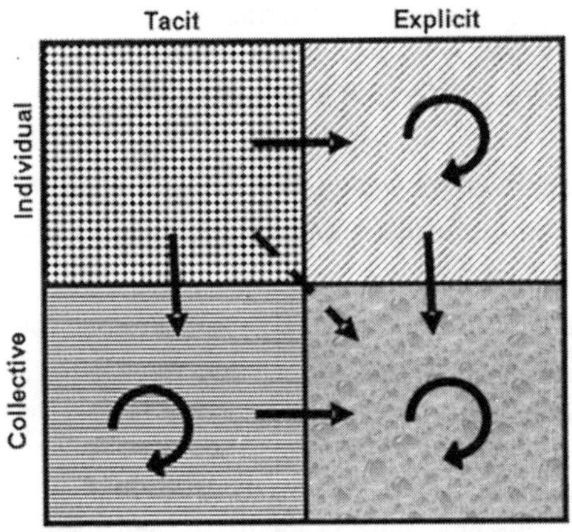

Figure 14.3. *The four levels of control for knowledge and different actions represented by arrows to go from one level to another*

14.2.3. *Actions from the Nonaka virtuous circle*

Thanks to the core knowledge cartography (Figure 14.2), we were able to carry out an action plan for the most critical knowledge. These actions were selected (according to their current level of control) among one of the four groups of actions defined on the Nonaka virtuous circle (Figure 14.4).

– direct transfer for tacit knowledge that will remain tacit (interesting for knowledge with very fast obsolescence, for example): to build up expert yellow pages, to develop co-operative work, etc.;

– indirect transfer of knowledge:

- explicitation (to make knowledge explicit) of tacit knowledge (which allows knowledge to be more timeless for the company, more perennial): writing of knowledge books, implementation of experience feedback process, etc.;

- sharing of explicit knowledge by the means of technology: creation of documentary portals, research engines, etc.;

- appropriation of knowledge: training, use of "information mapping" methodology, etc.

It should be noted that for an indirect transfer of knowledge, the three actions of increasing explicitness, sharing and then appropriation are essential. If one of them is missing, the transfer will be difficult. Software that claims to be KM tools is actually mainly based on sharing of explicit knowledge. If increasing explicitness and appropriation are not properly taken into account in a company, a document sharing tool will not be very effective for KM.

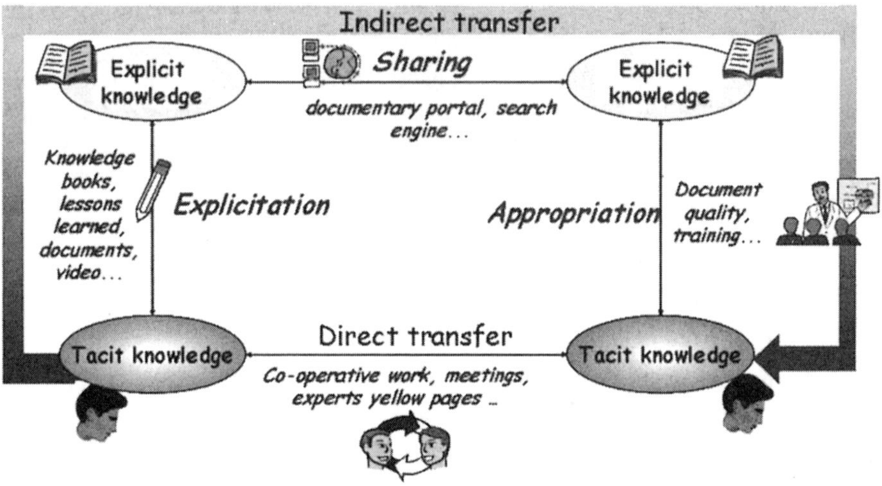

Figure 14.4. *The Nonaka virtuous circle representing different KM actions*

14.2.4. *A well appreciated approach*

Core knowledge cartography allowed us to build a capitalization and sharing action plan over one year. This action plan was based on a structured approach and not only on the predispositions of people or on their availability to work on KM. An annual updating of the cartography would enable us to follow up the progress of the capitalization work. Cartography is a very good tool and not only for the knowledge manager. For instance, we can use core knowledge cartography to build a training program.

A rigorous approach and an appropriation simplified by graphs make core knowledge cartography a good tool for communication and discussion. In a few minutes, it is possible to explain the definition of KM (with its various actions via the Nonaka circle) as well as the choices made for actions. KM is no longer just at the theory stage (valid for all companies), but it is now applied to the whole studied process.

14.3. Implementation of lessons learned

In order to avoid making the same mistakes again and to spread best practices, we implemented a lessons-learned capitalization process: instantaneous observations of capitalization (IOC). This process is based on two fundamental rules:

– experience feedback requires a treatment, an analysis by representatives of the various jobs involved before integration into a capitalization mode;

– filling in a lessons-learned sheet is never the end. It is necessary to define to users the capitalization mode that is the nearest to them. The aim is not to create new means for knowledge access but to complete or to improve existing ways. The capitalization action will never be to add a feedback card to a bank of cards but it can be: to modify a check list, to improve a generic risks analysis (FMEA) or to launch the writing of a knowledge book, etc.

The steps in the lessons-learned process are:

1. A card is filled in by a writer with reference to a model card.

2. An analysis of the card is carried out, as well as some interviews with people involved in this lesson learned (knowledge manager work).

3. A discussion of this card is held by a group (representatives of the various jobs), building up of a capitalization action plan with an associated leader.

4. Feedback to the writer to explain what is being done to prevent the problem from reappearing.

5. When the action is finished, validation of this one by the capitalization group and letting the writer know.

In eight months of existence, the IOC process gathered about 60 contributions, which is very close to the objective of one card per person per year. The IOC system is now part of everyday life. One of the success keys lies in the speed and simplicity for the writer. Five minutes are enough to fill out a card. One step which must not be neglected is to inform the initial writer of the capitalization progress. Moreover, steps in the process were already well known by employees because they are the same steps as used in a system already known for years: Instantaneous Observations for Safety (IOS).

At MHFR, the desire not to see a problem reappearing is stronger than the feeling of being dispossessed of knowledge. The employees were petitioning for such a process and now they are using it.

14.4. Knowledge explicitation

Many documents exist at MHFR. Communication is very often written because written documents are needed so that projects can be discussed with customers, suppliers and inside the company. However, at the moment, these documents are often dispersed and represent only one part of the overall vision on a subject. This is particularly obvious when the intensification of exchanges by e-mails is considered.

In fact, knowledge explicitation at MHFR consists of connecting the documents (or the knowledge they contain), like a spider's web, and completing them with a few missing explanations. After several researches and tests, we decided to use two tools in agreement with Information Mapping theories:

– MASK methodology and its graphic models, which enable the knowledge manager to analyze and collect knowledge through different points of view in order to build knowledge books;

– Mind Manager software (MINDJET), which makes it possible to create trees very easily (knowledge trees and expert trees, but also trees of concepts) and to export them in HTML format.

Thanks to the core knowledge cartography, several fields were found to have priority for explicitation. We started our work of explicitation by writing knowledge books on these fields. Depending on the kind of knowledge field (knowledge on products, on production processes, on transversal expertise, etc.), the shape of the knowledge book will be different. For example, a knowledge book on a production process will be based on the activities to set up this process, whereas a book on a product family is based on the functions that this product has to fulfill. So far, we have created four books at MHFR, three on product families and another on a manufacturing process.

The problems we have to handle are those met by the majority of companies that wish to clarify knowledge. The problem of updating these books needs to be considered along with the amount of time needed by experts to create the books. In the same way, MASK methodology still remains too complicated to be used by all experts. It requires too many efforts at model appropriation. So far, this tool remains in the hands of the knowledge manager.

In addition, clarifying documents that would be valid for the whole Mann+Hummel group involves other issues: the language (a language different from the mother tongue disturbs the appropriation) and the validation of the documents. Reaching agreement between experts inside an entity is very complicated but reaching a consensus between experts from several countries is even more complicated.

14.5. Sharing of explicit knowledge

At Mann+Hummel Automotive France, we used the cartography to improve the classification of our timeless technical documents. Indeed, in the absence of a documentary tool, documents have an annoying tendency to get lost in the maze of the Windows servers. It is sometimes difficult to find a document, and it is even more difficult to know whether this document exists. For the knowledge documents, we thus used the cartography to create a tree structure of folders representing the knowledge fields (through a Windows server).

Thanks to this tree structure, it is now easier to know that a knowledge document exists for a field and to find it but the system is not perfect:

– the knowledge tree requires a rather important time for appropriation if it is not used daily;

– documents (which generally present several knowledge fields) can be classified under several folders. A multidimensional classification would be really interesting.

Two families of documents exist:

– the knowledge documents, which are timeless;

– the projects documents, which have a limited lifespan, are generally very specific and have an applicability limited to the project in question.

The knowledge documents often come from project documents. In practice, projects do enrich our knowledge. These two families of documents are closely linked and sometimes it is difficult to differentiate them. It is thus important to have one common tool to manage and to have access to these both families of documents.

At MHFR, a study is ongoing to list the existing tools, the tools that will be available (coming from the Mann+Hummel group) and the functionalities necessary to meet this double need.

14.6. Direct transfer of knowledge

After the cartography and its translation into a tree structure of folders, we used the knowledge tree to create a simplified expert directory. For each folder (or knowledge field), we appointed a person responsible for the quality of this folder, both the content (to validate documents) and the form (to keep the folder clean). An HTML file representing the knowledge tree is used as an expert yellow pages (Figure 14.2). A person wanting more information on a knowledge field can contact the expert appointed for this field.

Thanks to its human size, this simplified expert directory is sufficient for search inside MHFR. On the other hand, it is not sufficient to locate an expert in a different firm or in a big entity. The need is especially felt between the various entities of the group. Work is on going to build a Mann+Hummel Experts Yellow Pages in order to identify an expert who could answer a problem. Cartography will help us for the identification of the experts and their assignment to the knowledge fields. The aim is to overcome the difficulties related to the different organizations of the entities. The interfaces and the research modes or navigation are still to be defined.

Moreover, discussion forums have been installed on the Intranet in order to openly ask questions to a group of experts and to enlarge communication modes.

14.7. KM: an everyday task based on people more than on technology

KM at MHFR consists of giving means of sharing experience to people. So far, we have improved organization and implemented methods for capitalizing knowledge, but we have not purchase any software. Software could be very helpful for developing co-operative work, workflow, document sharing, and we may invest in such tools in the future. However, technology is not sufficient for KM, the success of which is strongly linked to people involvement.

The role of the knowledge manager is really important. Part of his or her time is allocated to assisting people in capitalizing and in communicating and in following up actions such as the lessons-learned process. Without a knowledge manager, this kind of process would quickly disappear.

KM also requires employees' willingness to capitalize. A person not willing to share his or her experience cannot be forced to do so whatever the tools or organization are. It is part of the company culture. At MHFR, the strong sharing culture represents an important advantage for succeeding in a knowledge management approach. After two years of implementation, KM is now part of everyday work for MHFR employees.

Chapter 15

Thales System Engineering Community of Practice: A Knowledge Management Approach

15.1. Introduction

Thales (formerly Thomson-CSF) is a high tech company of more than 65,000 employees, established in 50 countries. It has three very different areas of activity: defense, aeronautics, and information technologies and services. Each area is split into "Business Groups". For example, "Defence", with 31,200 employees is now split into five "Business Groups" dedicated to different operational activities: naval, optronics, airborne systems, air defense systems, and communications. More than 30 civil and defense business domains are targeted: air traffic management, airborne systems, naval systems, geopositioning solutions, communications, C3I systems, security systems, simulation, etc., and 18,000 people are involved in Research & Development.

Three cross-company and entities have been set up to support and disseminate a common Thales culture: Thales Research and Technology (TRT), Thales University, and Thales Intellectual Property (Figure 15.1). The mission of TRT is to ensure that each worldwide unit has access to Thales knowledge in each discipline: methods, technologies and practices available and to be used.

Topics covered include program management, systems engineering, customer support, process improvement (CMMI), hardware and software. In this chapter we shall highlight the way in which a formal Knowledge Management (KM) initiative

---

Chapter written by Cécile DECAMPS and Michel GALINIER, Thales TRT.

helped system engineers in their day-to-day tasks. We are convinced that such a support is more and more mandatory for a company for which one the main added value is system design and engineering practices.

15.2. The knowledge sharing approach

"Knowledge sharing" is one of the five key values defined in 2000 by the Thales Board. It is the central element of the KM strategy within Thales and has the objectives of reduction in development cycle time (through re-use), mastery of technical choices (reduction of the risks), access to competencies and to expertise, help in mobility, and acceleration of innovation.

Knowledge sharing is an old story in Thales (formerly Thomson-CSF). It has been organized through cross-company subject communities, each based on a particular skill (Software (7,000 people), Hardware (3,000 people), Systems Engineering (2,500 people) and Microwave (300 people)) or based on common practices (Program Management (2,400 people), Process Improvement, Customer Services, etc.). These cross-company networks are called CET (Common Efficiency Teams). The first (Software and Hardware engineering) were launched in 1991. The objectives of each CET are defined by a Steering Committee, on which every operational unit has a representative. The purpose of these networks is to improve collective efficiency, to propose competitive solutions to the Business Units, to have a global vision of the community achievements and needs, and to manage the cross-company actions.

A CET concerns the self-management of a discipline by the actors in this discipline. With this worldwide approach in a given field, all the engineers share the same technical vocabulary and the same engineering process: they are ready to work together on a multi-unit project using the same basic discipline reference.

The knowledge management approach chosen by Thales is thus the result of two factors: on the one hand, strong management input ("sharing key knowledge values"), while on the other hand the existence of skill communities, already supported and organized in the CETs. Already involved in sharing practices, these communities, which have very numerous competencies and very high levels of expertise, very quickly showed their interest and motivation in going further with this approach.

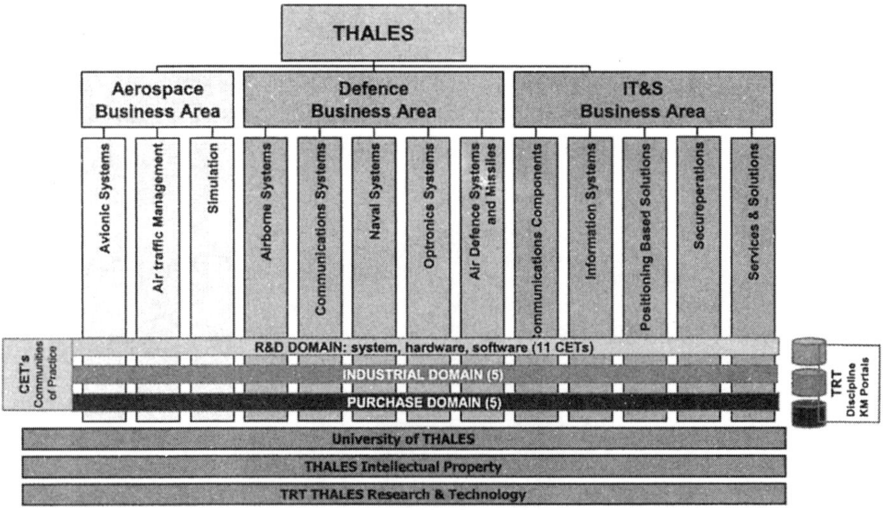

Figure 15.1. *The business areas and the cross-company communities of practice in Thales*

15.3. The Systems Engineering Community of practice at Thales

The Systems Engineering Community of practice is focused on three topics, processes, people and technology and three topic areas have been defined (Sys-EM and Sys-IE for technology and Sys-PI for the processes – Figure 15.2). Each area defines the granularity or the modularity of deployment in terms of rationale, products, services provided, delivery/support and acquisition process, and each of them is managed by a Technical Committee.

Systems Engineering Methodology (Sys-EM) leads the field: (i) by identifying the supporting system engineering processes; (ii) by defining the expected practices. Sys-EM is based on EIA 632 and CMMI standards.

Through the CMMI model, Systems Process Improvement (Sys-PI) provides assessment of practices (putting Sys-EM or any other methodology into practice), the roadmap for an improvement plan and process performance metrics.

Sys-IE: Systems Integrated Environment (Sys-IE) is concerned with the tools (COTS) and methods that support the processes defined by Sys-EM. Sys-IE develops incrementally according to the priority of needs and the available funding. Each increment provides the "how" or "what"; e.g. requirements management with DOORS, systems architecture with Rhapsody.

In terms of people management, the Systems Engineering community of practice acts as the "prime contractor" in relation to Thales University: which kind of training it is necessary to put in place to fit project needs and how many attendees can be predicted. This community has also proposed a Systems Engineering jobs classification to the Corporate Human Resources Department.

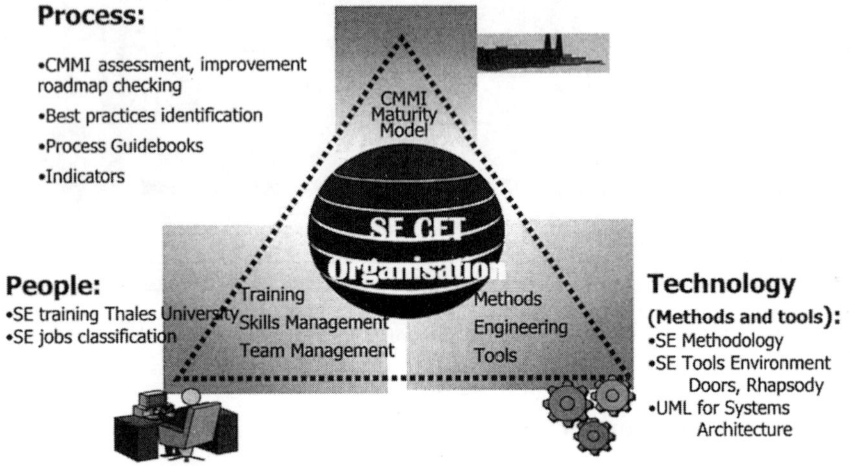

Figure 15.2. *Systems Engineering Community areas of interest*

The Systems Engineering Community is managed by a Steering Committee. Technical Committees, Working Groups and Shared Projects across the Business Units propose and implement solutions to address the actions and issues raised by the Steering Committee. A new rationale has been implemented: absolute majority rule has been replaced by relative majority rule. The objective is to speed up innovation and change management.

15.4. Why is there a KM portal dedicated to the Systems Engineering Community?

To support all these cross-company activities and collaborative projects, it has been decided to implement in Thales a complete KM approach so as to put into the hands of the communities dedicated tools (portal, groupware, etc.) and associated methodologies, support and organization.

One of the first communities chosen in the KM platform roll out process was the "Systems Engineering" community. It brings together more than 3,000 people throughout the world and has a very active CET. It also has the will to take a major step in the field of KM and to begin in-depth actions. This community was already mature from many points of view:

– It had many years of experience in the sharing of methods, tools, technologies and processes.

– It already had available CMMI ("Capacity Maturity Model Integrated"), a model developed by the Software Engineering Institute, defining the best practices in the development and the maintenance of systems and software and characterized by five levels, realizing the capacity of the company to master its development processes.

– Collaboration with the "Business Groups" was well established.

– It already had a very well organized intranet site.

The conditions were thus appropriate for optimization of knowledge sharing in this community, and a decision was taken to organize collaborative working spaces, allowing easier interaction, and to create knowledge bases that give fast and easy access to validated documents, to information and to expertise. Technologically, these two kinds of space are available via the "Systems Engineering portal" (Figure 15.3).

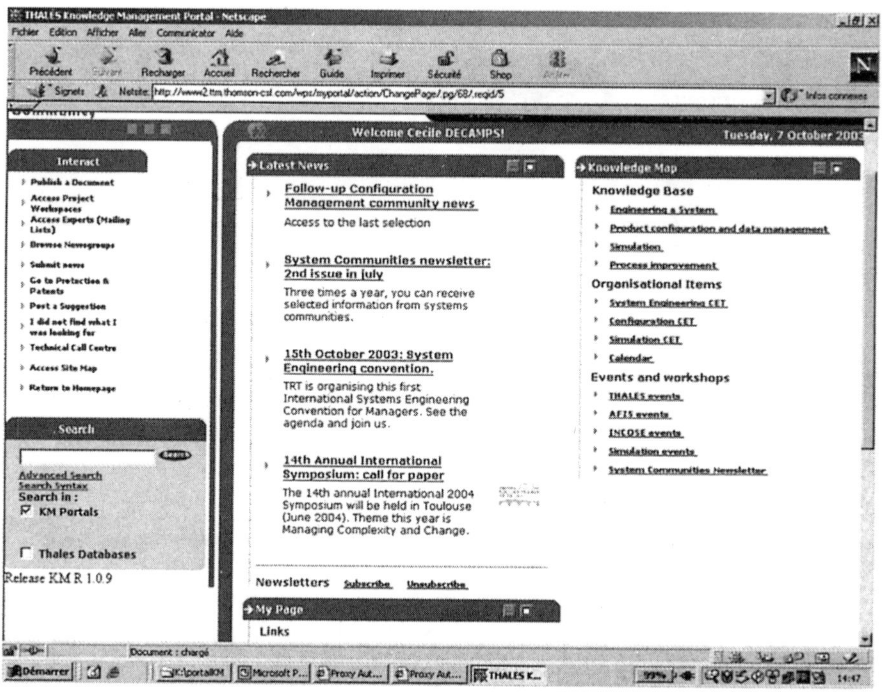

Figure 15.3. *Systems Engineering portal home page*

15.5. The Systems Engineering portal

The objective of constructing such a portal (as a focal point for accessing to knowledge bases and collaborative spaces) was to set up a systematic process to create, collect, synthesize, learn and use information and good practices and to more easily reach operational objectives.

Figure 15.4. *KM capabilities provided to each community of practice (product line)*

Three main families of functionalities have been set up (Figure 15.4):

– functionalities supporting SE community animation and management;

– functionalities supporting access to SE community people and competencies;

– functionalities supporting access to documents and other kinds of content.

15.5.1. *Functionalities supporting SE community animation and management*

Such a community cannot be "self-managed". It requires continuous and strong involvement to keep it active. The following KM tools bring their contribution to spreading and leveraging all management/coordination actions:

– *News*: a space dedicated to SE community news. News items can be internal (e.g. the new release of a recommended tool) or external (e.g. information about a competitor, a new regulation, technology watch).

– *Events*: the same range of ideas, internal or external events has to be shared as a necessary complement to "virtual sharing". In most communities, these more traditional events are very often the "starters" for new initiatives, which will then be pursued through virtual working groups, by using collaborative tools.

15.5.2. *Functionalities supporting access to SE community people and competencies*

The higher the "system dimension", the higher the level of expertise sharing should be. Being a system manufacturer and a prime contractor, Thales has to deal with more and more technologies and integration issues are more and more critical. So it is fundamental to organize in the most efficient way all kinds of connections/interfaces between experts in the different areas of systems engineering:

– *My experts*: this function highlights particular experts and their associated competencies. Experts help users in appropriation of existing tools and methods, spread best practices within Thales Group, etc.

– *My newsgroups*: user feedback and contributions are mandatory to enrich methods and further improve existing tools. Thematic newsgroups are thus available through the portal, which also acts as a discussion forum.

15.5.3. *Functionalities supporting access to documents and other kinds of content*

Knowledge base: this space is completely customized to meet community needs. This is the site for "core knowledge", available for both reading and publishing. For the systems engineering community, a first level of classification gives a "window" on the different sub-segments of the Systems Engineering community:

– engineering a system (systems engineering "core" issues);

– configuration management;

– simulation/modeling, etc.

Such a classification gives a first idea to the end user about the resources (human or otherwise) that can provide him with help.

The second classification level provides the user with the type of knowledge he will be able to access (Figure 15.5).

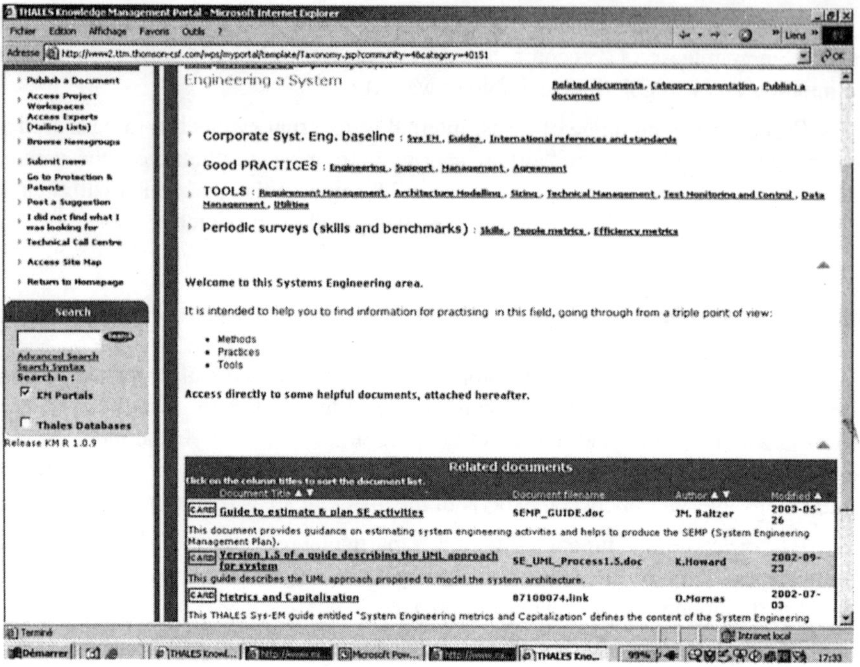

Figure 15.5. *Second classification level of the Systems Engineering portal*

All this information (best practices, guides, examples, etc.) will help users to obtain the maximum number of elements to support their operational needs. This classification has been carefully built by matching user needs (about 30 interviews with operational users based on samples of the Systems Engineering population) and "available" knowledge (what the community is able to propose through the portal).

– *My links*: in this area, the user has the possibility of customizing a part of the page according to his or her own operational needs:

- specific links ("favorites");
- shortcuts to specific executables or tools.

– *My workspaces*: the user has in this area access to all collaborative spaces in which he has been invited to participate (task forces, interests groups, temporary working groups). He will find in those spaces reports and working documents.

15.6. Rolling out/organization

In each community, the knowledge base of the portal has to be carefully set up and maintained. The knowledge assets are specific to the community and the portal is created by and for the community. It is not thus possible to make it live without the active participation of the whole community.

The evolution and the maintenance of the knowledge base cannot thus be a simple support activity. It requires the definition and distribution of precise roles through the community (KM roles):

– the "knowledge manager", who is responsible for the global quality of the portal and for appropriation of knowledge management processes by the community;

– the "knowledge integrators" (generally experts who validate the information that is published in the knowledge base);

– the "knowledge publisher", who publishes the information in the portal;

– the standard user, who can read all the information, but who can also propose a document for publication. The whole community is naturally concerned in this role.

15.7. The associated support organization

To support all these processes and organizations, Thales TRT has set up a four-person KM department, which is an important engine for the roll-out of such an approach. It realizes the essential work of project ownership and puts the platform at the disposal of the cross-company communities; it supports the maintenance and exploitation. This department also ensures that there is support for the content elaboration of the "knowledge base", in collaboration with the community experts and a few end-users.

The knowledge capitalization is not simply collection and storage of documents and information. It is a question of setting up a process of analysis, "explicitation" (making relevant) and structuring of the available knowledge in the community, whether it exists in an explicit form (documents, guides) or tacitly (experiences, examples). Integrating the different steps of contribution and re-use into the operational processes is also a key issue.

Access to connected disciplines (program management, customer services, process improvement, software, etc.) is of course mandatory for any systems engineer. This is also made possible through the access to similar portals, and through the possibility of making a document visible from several portals.

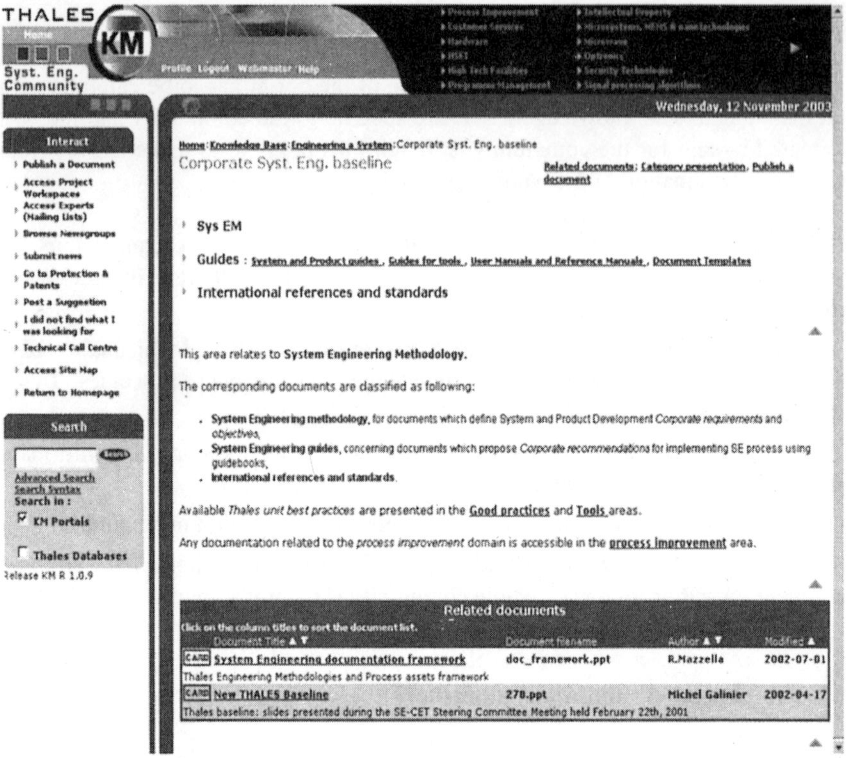

Figure 15.6. *Third classification level of Systems Engineering portal*

15.8. Balance and perspectives

A little under a thousand users from the Systems Engineering community are registered in the system engineering portal as we write this chapter. This good result highlights the operational need for such a portal.

Some work is planned to improve the user-friendliness character of the classification built for the knowledge base. This classification is the interface between the knowledge content and the user, and it must be very clear. A thesaurus dedicated to Systems Engineering activities is also being constructed to ease indexing/publication by all potential contributors.

Eleven portals were opened between June 2002 and October 2003.

The next objective is to increase the integration of these portals by having them more and more "fully integrated" in users' workstations, merged with other "information" channels available on the desktop (corporate information, project/program specific environment, Business Unit specific data, etc.).

The lessons of this project are numerous, but we can highlight a few elements that have contributed to making it successful:

– In term of strategy:

– ensure management commitment and support;

– build the device on a favorable ground: already formed communities, strong operational demand, motivated people;

– make the knowledge accessible by every person who is likely to re-use it or to make a contribution;

– consider the knowledge as an asset of the company.

– In term of people:

– create the culture of contribution/re-utilization;

– set up KM organizations (roles/sponsors, etc.).

– In term of process:

– integrate the stages of contribution and re-use into the operational processes;

– define standard processes to manage the knowledge capital.

Community managers now perceive KM roll-out process as a management tool, very structured in order to refine the objectives, to strengthen the community and to motivate people. The construction of a such a portal certainly requires fundamental questions such as the following to be raised. Who are my customers? Which are my products? Who are those I wish to make visible?

15.9. Bibliography

Beshore D.G., *Self-Organizing Maps (SOM) for Systems, Software Management and CMMI document Relevancy*, 13th Annual International Symposium INCOSE, 2003.

Davenport Thomas T.H., Probst Gilbert J.B., *Knowledge Management Case Book: Siemens Best Practice*, Wiley, 2002.

Dubain G., *Multi Disciplines Process Improvement in THALES using CMMI, 3rd European S-E Conference*, 21-24 May 2002.

Rus I., Limdvall M., "Knowledge Management in Software Engineering", *IEEE Software*, vol. 19, Issue 3, pp. 26-38, May-June 2002.

Chapter 16

Appraising the Knowledge in a Radio-pharmacy Center based on Process Mapping and Knowledge Domain Cartography

16.1. Introduction

Knowledge is being widely recognized as one of the most important assets, if not the most important asset, of any modern organization. As a consequence, this issue has also captured the attention of research and development centers, which are institutions whose missions depend on the creation and use of scientific and technological knowledge. The concept of treating organizational knowledge as a valuable strategic asset has become almost a consensus in these organizations and as such they must effectively create, capture, harvest, share, apply, preserve and protect their knowledge. Although the concept is clear for the majority of such centers, the methodology, means and tools to deal with the above-mentioned processes may not be in place, and may even not be in the awareness of the people in charge of managing those institutions.

Chapter written by Rita Izabel RICCIARDI and Antonio Carlos DE OLIVEIRA BARROSO.

16.2. The importance of knowledge identification and evaluation within organizations

As the ability to manage knowledge seems to become more and more necessary, it is recognized that the management of resources is effective only when this can be defined in terms of their relevance, demand, availability, repositories and other pertinent characteristics that can influence the administration of the resources. To achieve this, we need to map the relevant pieces of knowledge and to identify and characterize their attributes and repositories.

Most of the knowledge that is relevant for an organization is already inside its boundaries. It is organized (or scattered) within its systems, databases and files, and, possibly, a part of it is already incorporated into automated processes, although a substantial part resides within the minds and intrinsic abilities of the employees. Nevertheless, the contents of these repositories need to be constantly complemented with new knowledge, which can be acquired externally or generated in the organization, recycled, adapted and re-used in new circumstances. In summary, one can say that this core knowledge should be continuously leveraged and managed to generate more and more value with its use.

In principle, knowledge can be appraised with respect to its "criticality" by assessing its relevance (in terms of value aggregation to the organization) and vulnerability (possibility of loss and recovery difficulty). Furthermore, by evaluating the current maturity or development level of "pieces" of knowledge within the organization, and what would be ideal (practicable), a diagnosis of the knowledge gaps that the institution should cover can be produced.

Such evaluation is also necessary in order to have a coherent Knowledge Management (KM) strategy that guarantees the continuity and the development of this resource according to the objectives and goals of the organization, in terms of its vision of future. Knowledge identification and evaluation combined with the intended objectives of the KM help to unveil possible solutions that correspond to the organization's needs for each knowledge domain (capitalization, preservation, sharing, appropriation and knowledge creation) and it helps the prioritization of different KM initiatives.

16.3. The case study

16.3.1. *History and context of the Radio-pharmacy Center*

IPEN – the Energy and Nuclear Research Institute – is the largest research institute of the Brazilian National Nuclear Energy Commission (CNEN). It has

1,200 employees, 65% of whom are researchers or engineers. Brazil has developed a considerable amount of knowledge in the nuclear area over the last 40 years, through long-term investment in research and technology transfer. At present, as in many other countries, the nuclear field is undergoing long-term problems due to lack of research funds, low support from governmental policies and little interest from young students. As a consequence, this considerable body of knowledge is, to a certain extent, "decaying" as its rate of utilization and dissemination is diminishing continuously. Moreover, it is not being adequately renewed, since both the investment and the number of people involved in related Research & Development is also being reduced. Of course, there are some exceptions in a few particular application areas, but in general there is a large risk of non-preservation.

The Radio-pharmacy Center (CR) has been created by the transformation of a typical research unit, inside IPEN, into an industrial-like production unit with certain aspects of a business unit. It has been certified according to ISO 9001/2000 standards and has as its mission "to produce and to distribute radio pharmaceutical products for nuclear medicine (diagnosis and therapy)". Today it is the most important unit of IPEN from a social and economic point of view, since it supplies some 300 hospitals and nuclear medicine clinics in Brazil, covering about 98% of the demand for radio-pharmaceuticals in the country.

Geographically, 64% of the demand is concentrated in the south-eastern region, 14% in each of the southern and north-eastern regions, and 6% and 2% respectively in the center-west and northern regions of Brazil.

It is necessary to note that: (a) the Center employees are workers in the public sector, where there is limited management autonomy; (b) production of radioisotopes and radio-pharmaceuticals is still a monopoly of the government in Brazil; and (c) demand grows at 10% per year but nevertheless there has never been a problem of demand not being met.

Figure 16.1 shows the organization chart of IPEN with its units, where one can see the position of CR within it. Some of these units interact with CR rendering administrative, infrastructure and quality assurance support: administrative and information services, personnel training, etc.; infrastructure supporting services, etc.

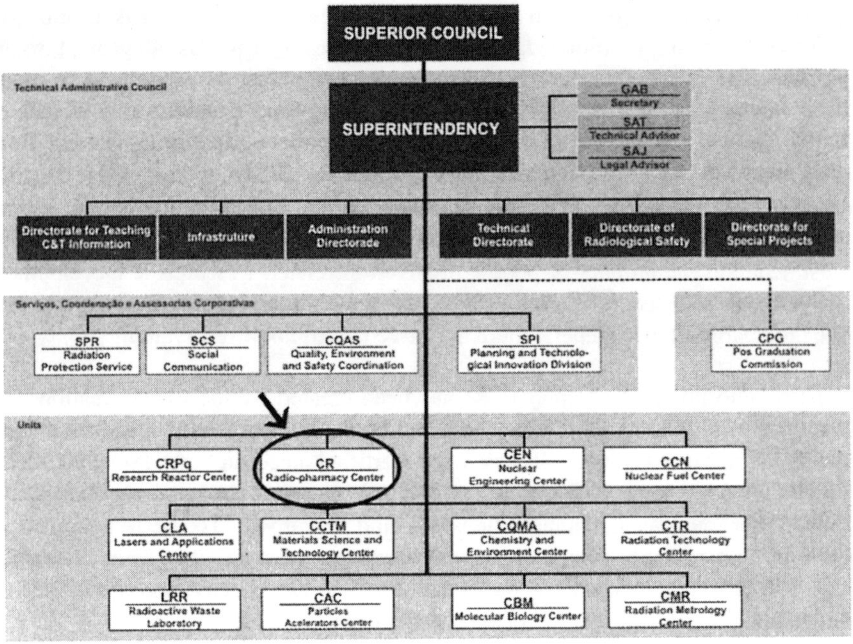

Figure 16.1. *The organization chart of IPEN and its units (taken from [INF 04])*

The CR management is organized in three divisions: Production, Quality Control and Assurance, and Research & Development of New Products (CRP, CRQ and CRPD respectively). The CRP division is subdivided into workgroups: Primary Radioisotopes and 99m Tc (technetium-99m) Generator, Labeled Compounds, Lyophilized Reagents (kits) and Support.

16.3.2. *The Center profile and key comments on its knowledge issues*

The history and initial objectives of the Radio-pharmacy Center have extensively influenced its current profile and its knowledge capital characteristics.

In the past, because of the mission of IPEN, the Center's activities were only geared to Research & Development in its field of interest (radioisotopes and radio-pharmaceuticals). The role of the staff was mainly to follow new discoveries in this field and to master its underlying processes, envisioning a future application of the outcoming technologies on behalf of the Brazilian community. For these reasons, most of the efforts were concentrated on obtaining new technologies from abroad and adapting them to the reality of the country. For any product, much research and experimentation was done and many prototypes were tested until it was proved possible to produce it in the Center. However, the efforts and the circumstances that allowed CR to migrate from its exclusive position as a research center to the current configuration of a business unit, scaling up the scientific research in laboratory to an industrial scale, were not preceded by any planning concerned with managing its intellectual patrimony.

In the following years, increased demand and the production of radio-pharmaceuticals with the degree of reliability required by clients caused the Center to adapt to its new functions and employees started to devote practically all their time to production activities.

More recently, with the growth of Center facilities, the increase in its production and the diversification of its products, it was necessary to enlarge the Center personnel. This involved the transfer of people from other areas of IPEN. However, many of those did not have the knowledge that was entirely appropriate to the needs of the Center, because, up to then, they had carried out activities in different fields. In addition, compared with CR people, they did not possess the extensive set of competencies and specific abilities, built through several years of research, because they had not participated in the activities that made up the development of the Center.

Today, however, it can be said that practically all the employees have the most important knowledge and abilities that are relevant to their functional performance. These were acquired partly as a result of their training or specialization and partly through accumulated experience in the development of their activities, contributing indeed to the importance of the intellectual capital of the Center.

If this historical background is analyzed and the primary vocation to research is kept in mind, it becomes clear that there has accumulated in the Center a vast knowledge repository of high technological content concerning the research and development phases of the current products. Many of the researchers who were responsible for most of the projects that have yielded this knowledge are still working in the Center. Also, some part of this knowledge has been structured and codified, because of the ISO certification. However, a very significant part of it still remains unstructured; some of this may be in explicit forms such as personal notes and scientific publications, but the largest part remains tacit in the mind of the researchers and engineers, some of them near retirement age. One can say that there is a significant portion of knowledge that is at a considerable risk of being lost, at least, partially. This characteristic is, to a certain extent, common to research areas that have not been concerned about management of their knowledge assets, even if they have become enterprising units at a given moment in their history.

Most of the processes developed, involving production, quality control and radiation protection, are structured in the Operational Procedures and Work Instructions of the Quality Management System and they are available at the operators' work place, either on-line or off-line. This set of documents constitutes a substantial part of an "Organizational Memory" (OM), the part that describes in a procedural form (operational processes) "what is done in CR" and "how it is done". However, the essential part of the "CR intelligence", in other words, the reason why a process is carried out in a particular way and how it has arrived at this present form, is not structured and registered. It is really important to include those portions of knowledge (CR intelligence) in the Organizational Memory (OM documentation). What we mean here is to register, in a didactic and objective way, the knowledge and the reasons why the current processes have reached their current form, as well as the reasons why the products in today's portfolio were developed, including which other technological routes were considered and tried, together with an explanation of the reasons for choosing one and abandoning the others.

It must be mentioned that there seem to be no barriers among the personnel to sharing knowledge. Through internal consultations, when a person has a need, he or she can usually find someone to share the desired knowledge, although this is usually sporadic and non-systematized behavior.

Besides its operational divisions, the Center has recently created a Research & and Development Division, which is in charge of research projects for the development of new radioisotopes and radio-pharmaceuticals. This division will also contribute to the development of improvements and modifications in the operational activities of CR, something that is currently under the control of some senior people from production.

A set of KM actions to accelerate the interaction between these two areas and, at the same time, to facilitate the systematized documentation and recovery of all generated knowledge would be of great importance. Such initiatives would transform the OM into a lively repository and a very useful instrument for organizational learning, as well as resulting in shorter time for the development of improvements and modifications.

This summarized diagnosis describes the main "frontiers" in CR, where KM would produce sensitive gains. In addition, other critical areas can certainly be pinpointed, whose processes could be improved and facilitated through KM. For such an aim, knowledge identification and evaluation are necessary to identifying these "frontiers" and to suggest the best KM actions that will close the gaps.

16.4. The KM project

A five-phase approach was adopted for this pilot project. First, an internal analysis was performed with the objective of identifying and understanding the main processes. Second, the enabling knowledge for these processes was identified and its repositories were characterized, both phases proposed by Barroso [BAR 01] and used by Silva [SIL 02]. Third, the identified knowledge was organized in a knowledge cartography representation. The fourth phase consisted of a "criticality" analysis according to a set of criteria chosen from a large collection suggested by the French KM Club [COL 00] and adapted to the context of the Center. The last phase used previous results to propose a set of sensible KM initiatives for the Center.

16.4.1. *Study of processes*

The study of the processes had the objective of more systematically understanding the foundations of the operational processes and their interfaces. Knowledge identification was accomplished through the study of center processes and underlying activities. Different means were used in a complementary mode: documentation review and analysis; interviews with leading personnel of the organization; writing up of summaries; and validation with the experts. Processes defined how the inputs work and how the existent resources are used so that the organization fulfills its mission.

In fact, this work was facilitated by the thorough documentation available as a result of the ISO certification. The Radio-pharmacy Center was studied in terms of the logic of its processes, their work and the information flow, as well as their interfaces. The processes of CR have been described in a classic manner using flow diagrams. The macro-processes were identified from the Quality Management System Documentation (Integrated Management Norms, Operational Procedures, Work Instructions and other documents) and they were decomposed into the respective processes and activities. From this analysis, the processes enabling knowledge were identified and characterized.

16.4.2. *Knowledge identification ("enabling knowledge")*

Once the Center processes were understood, a discussion with the (knowledge) actors of the processes helped to classify more precisely which knowledge and operational abilities are necessary and sufficient to achieve an adequate outcome from each process. This fine process analysis made it possible to elaborate and table related processes, activities and knowledge (with some supplementary information on products). To be thorough, in this pilot project, it was decided to focus on all production and research and development processes. Therefore only a few off-core processes were not included. Several tables for detailing processes, identifying their enabling knowledge, were elaborated, although, for reasons of confidentiality, this information could not be included here.

16.4.3. *Construction of the knowledge cartography*

An interesting problem, for which there is not a widely accepted solution, is how to represent the organizational knowledge in a way that produces a representation that is both visually friendly and accurate, in other words, how to have a good reflection of the organization's intellectual assets.

There are many ways to picture, in structured fashion, the tangible resources of an organization. However, knowledge assets are more difficult to represent. The approach used in this project, called "Knowledge Domain Cartography", is based on a classification by domains, according to themes and final outcomes, which was proposed by Ermine [ERM 02] and used by Peil *et al.* [PEI 01] and Aubertin *et al.* [AUB 03].

The construction of the cartography starts with a central node that corresponds to the main purpose of the organization. Then, a set of outward flowing axes starts from this node, each one representing a strategic knowledge theme, usually associated with one of the main components of the mission of the organization. Depending on the level of detail that one wishes to show, the main axes can have secondary axes representing sub-themes and these give rise to branches that represent the knowledge domains. Sometimes, these domains can even be further split into sub-domains. Usually a top-down approach is used, with some charts to show themes and sub-themes and then separate charts to detail each axis (theme) into sub-themes, domains and eventually sub-domains. Sometimes some centered support axes are used to represent important knowledge that is not directly connected with the main purpose of the organization but with support activities related to it. A more detailed description of the Critical Knowledge Domain Cartography representation was published by [AUB 03].

The Knowledge Domain Cartography of the Center was represented using eight diagrams. The first one gives a general view of the main axes, or strategic knowledge themes, according to the following categorization: Planning, Production Technology, Research & Development, Quality Control, Radiation Protection and Special Processes; plus a support axis (Norms and Regulations). The next seven are used to detail the domains pertaining to each of the main axes down to the level of domains or sub-domains.

Figure 16.2 shows the general vision of the Knowledge Domains Cartography of CR. To avoid a very dense drawing, only the sub-themes of Production Technology have their names shown.

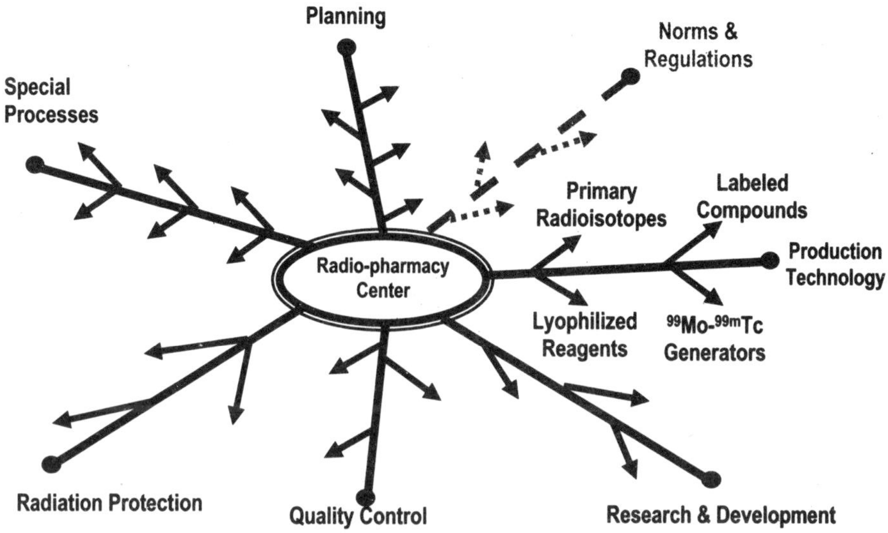

Figure 16.2. *Knowledge Domains Cartography of CR (general vision) (translation from [INF 04])*

16.4.4. *"Criticality" analysis*

The objective is to assess the relevance of each domain in terms of the objectives and goals of the organization and the degree of vulnerability of the objective. This kind of analysis gives important hints for choosing the most effective KM solution for each domain.

In this project, the analysis model created was based on the reference [CLUB 00], using evaluation criteria that were aligned with the goals and needs of the organization. Two criteria were used to "measure" the relevance of the knowledge domains – quality and complexity of knowledge and importance for the strategic objectives of the organization – and three criteria were chosen for the vulnerability assessment – difficulty of knowledge acquisition, capacity for sharing in the context of CR and knowledge rarity. A three-point scale was used for each criterion (0, 1.5 and 3).

Questionnaires and planned interviews were used for the analysis. The choice of the experts to collaborate in the assessment was based on their leadership, experience and proficiency in knowledge area of assessment.

The compound grade for relevance was an average of the grades for its individual criteria and the same was done for vulnerability. A knowledge domain was considered critical if the global grade was greater than or equal to 1.5 (grades are 0, 1.5 and 3) and if there was a grade 3 in at least one of the criteria. About 30% of domains ended up being classified as critical. The themes in the cartography of CR that were found to have critical knowledge domains were: Production Technology, Planning, Special Processes and Research & Development.

As an example, Table 16.1 shows the details of two sub-themes of the Production Technology theme. The knowledge domains and sub-domains are shown and those that were considered critical are identified.

Theme: ***Production Technology***			
Sub-themes	**Knowledge domains**	**Knowledge sub-domains**	**Critical knowledge domains**
Primary radioisotopes	Radioactive material processing	Cyclotron target preparation	X
		Reactor target preparation	X
	Supporting techniques	Glove-box handling	
		Glove-box cleaning	
	Radioisotope processing	Solutions preparation	
		Dilution calculations	
		Ph measurements	
		Impurity segregation	X
		Activity measurement	
Labeled compounds	Supporting techniques	Glove-box handing	
		Glove-box cleaning	
	Compound processing	Solutions preparation	
		Labeling	X
		Activity measurement	
		PH measurement	
		Dilution calculation	
		Purification	

Table 16.1. *Theme: Production Technology of the Knowledge Domains Cartography (translation from INF 04)*

16.4.5. *Setting up a KM plan of action*

There are many actions in the "classical" KM repertoire that can be suggested for leveraging generation, sharing, utilization and improvement of the knowledge domains that are considered to be essential for the Center. The best choices depend on many factors, such as the type of knowledge conversion, formatting and the actions appropriate in each knowledge domain.

The identification, mapping and analysis that were performed in this project provide a clear perception of which the critical knowledge domains are, how critical they are and why. This is the kind of knowledge that allowed a shortlist of tailored KM actions to be proposed, as follows:

– knowledge data (video and sound) concerning the critical aspects of the processes (critical theme: Production Technology);

– a knowledge database of "problems and solutions" (critical theme: Production Technology);

– a program "RC teaching RC" – meetings and mentoring (critical theme: Production Technology);

– virtual communities of practice (critical theme: Production Technology);

–an internal program of Contamination Control and Best Practices in Fabrication (critical theme: Special Processes);

– a KM-oriented functional development program (all themes);

– an organizational memory (critical theme: Research & Development);

– a knowledge portal (all themes).

A few initiatives are currently in progress, such as the elaboration of an organizational memory using knowledge books and the MASK method [ERM02].

16.5. References

[AUB 03] AUBERTIN G., BOUGHZALA I., ERMINE J.-L., *Cartographie des connaissances critiques: extraction des connaissances et apprentissage*, Hermès, 2003.

[BAR 01] BARROSO A.C.O., *Processos Corporativos e Outros* [lecture notes NA 5773, 2o. Sem. 2001].

[BAR 01] BARROSO A.C.O., *Alguns Conceitos e Motivação Pessoal.* Available at http://www.terraforum.com.br/sites/terraforum/biblioteca, accessed 2005.

[COL 00] Club Gestion des Connaissances, Ermine J.-L., Laude H., "A Knowledge Maturity Model", Proceedings of the "Knowledge Management; Theory and Practice" Workshop, J-L. Ermine (dir.), *PKDD2000 (Principles of Knowledge Discovery from Data)*, pp. 13-18, Lyon, 12 September 2000.

[ERM 00] ERMINE J.-L., *Les systèmes de connaissances*, Hermès, 2nd ed, Paris, 2000.

[ERM 02] ERMINE J.-L., *La gestion des connaissances*, Lavoisier & Hermès, Paris, 2002.

[ERM 02] ERMINE J.-L., *Initation à la méthode MASK*, CD-ROM edited by the University of Technology of Troyes, 2002.

[INF 03] *Informe Anual IPEN*, Publicação IPEN, August 2004.

[PEI 01] PEIL O., AUBERTIN G., ERMINE J.-L., MATTA N., "La cartographie des connaissances critiques, um outil de gestion stratégique des connaissances", Proceedings of the CITE'2001 Colloquium – Coopération, Innovation, Technologie; 29-30 November, 2001, pp. 237-308, Troyes, France.

[SIL 02] SILVA D.D. DA, SERNIC R., BARROSO A.C.O.; RICCIARDI R.I. "Projeto de gestão do conhecimento para uma grande empresa multinacional do ramo gráfico", In *XXII Simpósio de Gestão da Inovação Tecnológica*, 6-8 November 2002, Salvador, BA. Proceedings, 148 and CD-ROM.

Chapter 17

Case Study: Knowledge Preservation for a Nuclear Reactor

17.1. Introduction

The nuclear option was first introduced in Argentina in 1950. From then on until now, over half a century later, the Argentinean nuclear area has a hierarchy and a scientific and technological dimension that is important in all activities linked to the nuclear area. The export of research reactors, the production of nuclear fuels and the installation of nuclear power plants constitute some of the most outstanding achievements in this area.

In June 1974 the NPP Atucha I began commercial operation of the 313 MWe reactor (increased later to 357 MWe), the construction of which, through a turnkey contract, was awarded to Siemens-KWU. The reactor is a pressure recipient type using natural uranium (it now uses slightly enriched uranium) moderated with heavy water (HWR). In 1984, the Embalse Nuclear Power Plant of 600 MWe (type Candu) was connected to the net on a commercial basis and in 1981 the construction of a third NPP Atucha II of 745 MWe was begun and awarded to Siemens; it uses an Atucha I technology reactor. At present, it is under construction with 82% of the work completed.

The full operation of these plants[1] (Atucha I and Embalse) led to the acquisition of experience and to the achievement of a high standard in knowledge about the

---

Chapter written by Marta EPPENSTEIN.

1 The power generation company is Nucleoelectrica Argentina S.A. (NA-SA).

operation of nuclear power plants. To this should be added the necessary know-how to provide most part of the supplies for the plants.

17.1.1. *Atucha-type reactors*

The following factors should be taken into account.

First, as Siemens, the designer of the Atucha-type HWRs, has transferred its nuclear activity to Framaton ANP, Argentina must undertake knowledge preservation concerning this type of reactor, whether it is decided to extend the operating of Atucha I NPP, and/or if it is decided to end the construction of Atucha II.

A second reason for undertaking a knowledge preservation program is the aging and increasing retirement of personnel in the nuclear field, together with the small number of young people in nuclear-related disciplines at the universities.

This situation motivated CNEA[2] to implement a Knowledge Management (KM) system so as to preserve the knowledge that had been gained from the Atucha-type reactor in order to transmit this to following generations. The aim of the program is to extend the useful lifetime of Atucha I and to complete Atucha II, to convey this knowledge in real time, and to encourage creativity and innovation. The operation and maintenance tasks, as well as the reactor decommissioning and dismantling, will thus have a better access to the knowledge capital involved.

17.2. Practical approaches

17.2.1. *Strategy analysis*

The strategy involved recognizing the critical knowledge to be preserved. A methodology was used for its identification, which incorporates the map layout technique, related to reactor knowledge.

A knowledge map is a tool for structuring the capital knowledge of an area or domain. It is a comprehensive visualization of the domains of available knowledge in the company (or the organization or the service concerned); it should be like cartography [PEI 01].

The construction of a map or cartography is fundamentally intuitive. There is a road that leads to knowledge and an attempt is made to follow this road by using information content, experts and resources, which may or may not belong to the

2 CNEA, the Atomic Energy National Commission of Argentina.

organization. The map is not an inventory of things but a method of locating places and a tool that uses cognitive surfing in order to access the organization inherited knowledge of the organization, which may be tacit, if it indicates the people who possess knowledge, or explicit, when it indicates information sources. In other words, a map defines the knowledge capital. Management of the knowledge capital is a key factor for the organization and, to achieve this, a detailed analysis should be made with the purpose of determining what knowledge should persist, should be developed or should be given up.

17.2.1.1. *Different approaches to the map*

There are different approaches that can be used in organizing the cognitive resources of the company or organization, in this case the reactor:

– those based on the organizational structure;

– those based on the processes;

– those based on activities, topics or purposes.

In this particular case, the most appropriate way is to use the topic or domain of knowledge classification, since the organizational structures are likely to change through time and the one of processes is complex to develop and apply.

It is also possible to use another approach in a particular area, so as to make the map more comprehensible.

The Hishikagua-type diagram was used for the map representation. This diagram made it easy for us to access different levels as it represents the cognitive models of knowledge hierarchically (Figure 17.1)

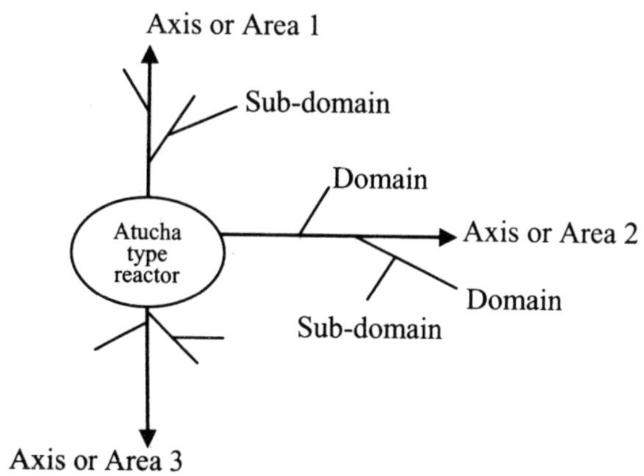

Figure 17.1 . *Knowledge map representation*

17.2.1.2. *Identification of knowledge axes*

Building the knowledge map requires a lot of effort, both by individuals or in group work by the specialists and experts as well as by the knowledge engineer.

The model must be accessible and practical. The representation should be adapted to the way in which it is likely to be used. As the construction proceeds, for those responsible for validating what has been created, it is an iterative process. Finally, the head of the organization must validate the map.

The key axes or areas, with their corresponding Atucha-type reactor domains and sub-domains, which represent the skills and specific experience through from design to operation, are as follows:

– Axis 1: Reactor design

– Axis 2: Reactor fuel

– Axis 3: Mechanical components and accessories

– Axis 4: Systems

– Axis 5: Safety

– Axis 6: Electric and Instrumentation and Control (I&C)

In addition, there is a transverse axis of support function tasks that are linked indirectly to the reactor.

The classification at which we have arrived has a huge overlap with the categories adopted by the INIS in its document: *Subject Categories and Scope Descriptions ETDE/INIS Joint Reference Series* No. 21. [IAEA 02]:

– Reactor theory and calculation (including experiments to verify calculations). Describes knowledge about the reactor itself (the physics and the thermohydraulics, the neutronic parameters, kinetics, etc.).

– Reactor components and accessories. Description of knowledge about the mechanical components, design, manufacturing, performance, maintenance, vigilance at service, etc.

– Reactor fuels. Everything related to design, manufacturing, validation tests, manipulation systems, etc. during the complete lifetime of the power plant.

– Reactor control systems. Control and instrumentation, protective systems for the reactor, scram's logics (the logic applied to produce an automatic reactor tip), etc.

– Liability for nuclear damage protection (security, radioprotection, rules, instrumentation related to radioprotection, accident analysis, risk analysis).

The documentation that was referred to also enabled us to establish a wider domain description, including the scope, and the limits and contour conditions of boundaries with other domains or components. It also facilitated the division into sub-domains and specific areas.

The description of the reactor map provides a taxonomy for the Atucha-type reactor, providing access to the subjects and scope of INIS for this specific reactor type.

17.2.2. *Identification of the critical knowledge: approaches*

A knowledge domain is critical because it needs to be capitalized or shared or it requires innovation. It is convenient to track down certain critical approaches linked to strategy of the organization, in order to classify them.

The establishment of approaches depends on the organization in which the analysis project is located. This depends on the environment of their activity, their strategy, their external and internal networks, their culture and their history, etc.

We made use of a library of approaches to evaluate the knowledge criticality. We have made an evaluation considering four possible thematic topics for the knowledge type [CLUB 03]. In the following sections, we give describe the critical knowledge analysis for each of the four types.

17.2.2.1. *Rare or unable of replacement*

– The company is the only one that holds this knowledge (leadership, originality, confidentiality) or this knowledge exists abroad (outside the country where the company is situated) and thus is not accessible.

– There are few people who hold this knowledge in the company and otherwise it is not accessible or available.

– It is not possible to replace this knowledge by other knowledge that will allow the company to obtain the same results.

17.2.2.2. *Usefulness for the company*

– This knowledge belongs to a key area or department in a company, which is considered strategic. This is because:

- it is closely connected to the mission and purposes of the company;

- all its component parts create value (customer satisfaction, beneficiaries, shareholder interest, employee interest, community interest);

- eventually an emergent area could be created, and it is likely to generate value over a longer time period.

– This knowledge can be reused in another area (reuse and adaptation of the domain to the necessities of another area).

17.2.2.3. *Difficult to obtain*

– It is difficult to identify the sources of the knowledge (people or documents).

– Access to the knowledge may imply the creation or the mobilization of contact networks.

– The kind of knowledge is basically tacit.

– The knowledge evolves quickly.

17.2.2.4. *Difficult to use*

– This knowledge is deep and complex, difficult to make appropriate.

– It is necessary to know the history of the area concerned area in order to understand and to integrate the knowledge.

– A proper knowledge of the environment and network (internal or external) is required to apply this.

The approaches were based on these groups, which then allowed us to evaluate degree of knowledge criticality.

The criticality analysis for each case was supplemented with the associated information, type of documents, related staff working at present or having worked in the past, databases, ICT architecture, etc. For each critical domain, depending on the reasons why it is critical, one can see what has been done and what can be done in order to improve the criticality.

Different KM actions that put a KM process into operation can be started up depending on why it is critical:

– capitalizing it by obtaining a document, databases, knowledge communities, etc.;

– sharing knowledge by means of forums, groupware, seminars, meetings, etc.;

– appropriation of the knowledge through learning, lessons learned;

– receiving information from outside the organization by means of e-business, recruiting revision, etc.

The result is an important knowledge group with a certain degree of criticality, which fits the approach type already analyzed. This will be presented in a future paper.

17.2.3. *Building the knowledge map*

We have now rather formally defined a powerful graphic tool for describing the reactor knowledge capital in a hierarchical way so that it can be preserved. Besides being used for the criticality analysis, it is also used as an access gateway to the knowledge capital, indicating, according to the knowledge area, people's skills, publications, corresponding documents and other information (Figure 17.2).

17.2.4. *Knowledge server*

The knowledge map marks the beginning of planning the management system. As a graphic interface, the knowledge map provides us with the knowledge server design. This will allow the user to browse through the capitalized reactor knowledge. The map domains are related, through hyperlinks, to several, very heterogeneous information sources: documents, publications, manuals, reports, events, experts' yellow pages, databases, software, etc. This kind of site is a knowledge key providing access to appropriate knowledge and information.

The use of the map to design a knowledge server is a project mainly related to the information system, requiring the necessary resources for its planning and development.

A significant effort has been made to make the critical data available on the knowledge server. Report tasks, recognition and possible homogenization of the document structure to identify the different systems and ways of data storage, types of documents and kinds of possible architectures have to be carried out in order to obtain a server or an intelligent portal. A great deal of important information related to the critical knowledge has been identified, for example, technical reports, drawings, registrations, manuals, operation instructions, events, design, etc. Certain related files, such as magnetic tapes, digital converted videos, microfilm, word processors in different versions and others, also have been analyzed.

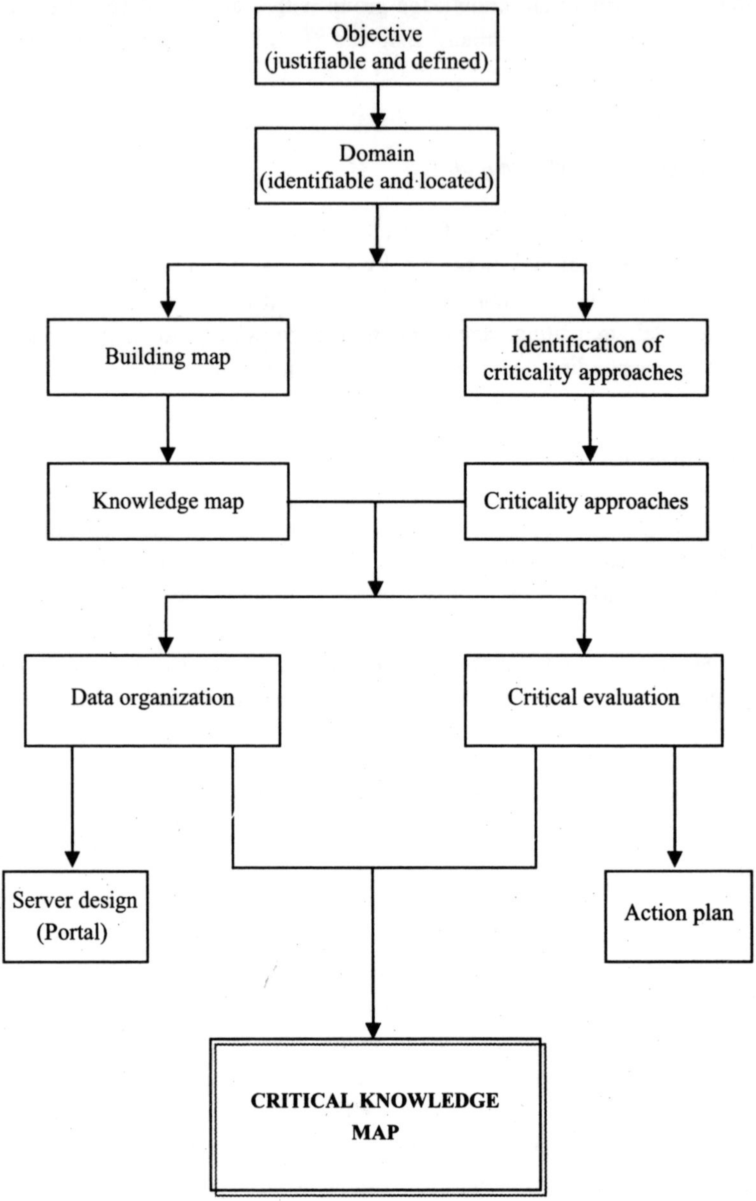

Figure 17.2. *Critical knowledge map building*

17.3. Technical development: analysis factors

17.3.1. *Knowledge transfer and capitalization*

The term "Knowledge Management" appeared in the last decade of the 20th century in big companies where the problem of processing information played an important role. It was essential for these companies to ensure a proper use of the information gathered over the years. Knowledge that has not been used for a certain period of time does not lead to the development of new ideas, whereas knowledge that is used is then capitalized and, if shared, also generates new knowledge.

The theory of Nonaka and Takeuchi [NON 95], which is called the knowledge cycle, describes these processes (capitalizing and sharing). There we can see how knowledge undergoes evolution from tacit to explicit and which sub-processes will take place; see Figure 17.3.

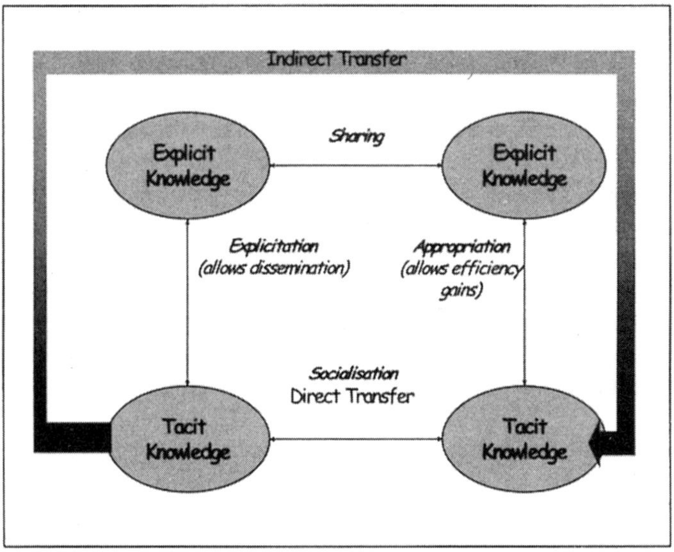

Figure 17.3. *Knowledge cycle*

The direct transfer, called socialization, is where tacit knowledge is shared through workshops and meetings among experts. The indirect transfer is another alternative for sharing knowledge:

– The first step is to make the tacit knowledge explicit. At this stage it is possible to carry out knowledge engineering in order to model the use of the correct methods.

An approach based on artificial intelligence, together with experts systems, is feasible. We use the MASK method (Method of analysis and structure of knowledge) developed by Ermine [ERM 02].

– The sub-process that follows consists of sharing the explicit knowledge by using information technologies such as the Intranet.

– Once the shared knowledge has been appropriated, it generates new knowledge through group work, lessons learned and training; this is the stage where the "organization learns".

The method applied for capitalizing the knowledge, as we have said, is MASK. This enables the representation of patterns that are recognized by the experts from whom the knowledge is elicited. It is a method that allows know-how to be extracted. This task is done by techniques based on meetings or interviews between the expert and the knowledge engineer. They build together the pattern that represents a certain process. The expert contributes the most valuable thing, which is his or her knowledge and the most important achievement for him or her is to interpret his or her own thought and be able to put it into words, gestures and other means of expression for its communication. The related information is analysed and it is possible to design a pattern recognized by the expert. It is this description of his or her subject (mirror effect) that has helped him or her to bring out and model what can be finally validated. The same approach can be used with several experts if the subject or domain requires it. In this way it is possible to get a consensus representation of the knowledge.

In relation to the project planning, it is important to consider the time that experts have available for developing the cognitive structure of the reactor domains. These experts are usually nuclear power plant operation and maintenance staff, fuel manufacturing staff or staff involved in another reactor-related activity.

Some procedures and guides have been written to improve project performance: on carrying out interviews, producing reports or notes, and use of related documentation and information, such as validating what has been done and all that that involve so that the pattern can be obtained in a systematic, ordered and efficient way. For each one of these experiences there are feasible learned lessons for the knowledge capitalization (LL_n).

17.3.2. *Human resources*

The integrated knowledge relating to the Atucha-type reactor includes human resources that belong to science and technology organizations with different cultures from (the knowledge of) the staff of nuclear power plants and other production plant staff; this may be associated with the reactor knowledge but have a more corporate culture. The way of knowledge sharing is different for the two types. However, they can share their goals in meetings and science seminars where they exchange and discuss specific topics. This is another challenge for reaching a level of critical knowledge as well as improving the criticality (LL).

By means of a technical analysis of the ages of the specialized personnel age in NPP and of nuclear field scientists in R&D at CNEA (a scientific–technological organization), it is clear that both organizations have the same characteristic. 50% of the staff are over 50 years old and many of them will be retiring in the next five or 10 years.

The information flow cultures, together with personnel retirement, are key analysis factors in defining the strategy for knowledge preservation. Two important matters must be kept in mind:

– the eventual loss of knowledge caused by experts' retirement;

– the generational change as a result of employment of new personnel.

The application of the techniques and methods that have been developed could be applied in each case to make the KM system richer and to ensure knowledge preservation.

17.4. Conclusion

The knowledge preservation project (KP) applied to the Atucha-type reactors used different kinds of techniques and methodologies to identify the critical knowledge domains relating to the reactor and identified their criticality by means of a knowledge map. This tool will enable us to apply KM processes and to design a knowledge portal.

Experience obtained through the development of the KM system shows how training techniques are put into practice, in order not to interfere with normal plant operation, and how KM processes can be initiated in order to improve the criticality.

17.5. References

[CLUB 03] CLUB GESTION DES CONNAISSANCES, *Bibliothèque de critères* Paris, 2003.

[ERM 02] ERMINE J.-L., *Initiation à la méthode MASK*, University of Technology, Troyes, 2002.

[IAEA 02] IAEA, *Subject Categories and Scope Descriptions*, ETDE/INIS Joint Reference Series no. 2, IAEA, Vienna, 2002.

[NON 95] NONAKA I., TAKEUICHI H., *The Knowledge Creating Company: How Japanese Companies Create the Dynamics of Innovation*, Oxford University Press, 1995.

[PEI 01] PEIL O., AUBERTIN G., ERMINE J.-L., MATTA N., *La cartographie des connaissances critiques, un outil de gestion stratégique des connaissances*, presented in the CITE'2001 Colloquium, University of Technology, Troyes, 2001.

17.6. Further reading

ERMINE J.-L., *La gestion des connaissances*, Lavoisier Hermès, Paris, 2003.

IAEA, *Managing Nuclear Knowledge*, IAEA, Vienna, October 2003.

The Authors

Gérard AUBERTIN
Walter BAETS
Jean-Marie BEZARD
Imed BOUGHZALA
Nigel COURTNEY
Cécile DECAMPS
Antonio Carlos DE OLIVEIRA BARROSO
Aurélie DUDEZERT
Patrick EPINGARD
Marta EPPENSTEIN
Jean-Louis ERMINE
Pierre FAYARD
Michel GALINIER
Chris HENDRY
Clive HOLTHMAN
Réal JACOB
Nathalie LE BRIS
Olivier LEPRETRE
Rita Izabel RICCIARDI
Jean-François TENDRON
Thierno TOUNKARA
Jean-Michel VIOLA

Index

A

B

C

D

E

F

H

I

K

M

P

S

T

V